FASHIONING AUSTRALIA'S FORESTS

Remembering the loving encouragement of
Alan Dargavel and Kathleen Dargavel

Fashioning Australia's Forests

JOHN DARGAVEL

Melbourne
OXFORD UNIVERSITY PRESS
Oxford Auckland New York

OXFORD UNIVERSITY PRESS AUSTRALIA

Oxford New York
Athens Auckland Bangkok Bombay
Calcutta Cape Town Dar es Salaam Delhi
Florence Hong Kong Istanbul Karachi
Kuala Lumpur Madras Madrid Melbourne
Mexico City Nairobi Paris Singapore
Taipei Tokyo Toronto

and associated companies in
Berlin Ibadan

OXFORD is a trade mark of Oxford University Press

First published 1995

National Library of Australia
Cataloguing-in-Publication data:
Dargavel, John.
Fashioning Australia's forests.

Bibliography.
Includes index.
ISBN 0 19 553526 X.

1. Forests and forestry – Australia – History. 2. Forest conservation – Australia – History. 3. Forest policy – Australia – History. I. Title.

333.750994

Edited by Janet Mackenzie
Cover design by R. T. J. Klinkhamer
Text designed by R. T. J. Klinkhamer
Typeset by Superskill
Printed by Bookpac Production Services, Singapore
Published by Oxford University Press,
253 Normanby Road, South Melbourne, Australia

Cover painting by W. Lister Lister, Australia (1859–1944), *Loading the Jinker*, 1914, detail, oil on canvas, 183 × 231 cm, Art Gallery of New South Wales.

CONTENTS

ILLUSTRATIONS

Plates

Figures

Tables

PREFACE

The passionate contests between those who would fell Australia's native forests and those who would leave them be provide revealing examples of wider problems. Economically, Australia is struggling for its place in the rapidly changing economy of the Pacific Basin. Environmentally, it, like the whole world, has to find how a sustainable future might be created. Scientifically, it seeks the knowledge to manage the nation's resources to both economic and environmental ends. Socially, many people need a more equitable society, while others strive to make it less so. And politically, governments have to deal with the changes, crises, contests and conflicts that these imperatives create.

Australia's forests have been fashioned, since 1788, by changing, disparate and often contradictory economic and political interests – imperial and colonial, international and national – enmeshed with changing ideas and moral claims of conservation, development, decentralisation and now environmentalism. As the state still retains ownership of three-quarters of Australia's forest land, many of the contests about how, and in whose interest, the forests should be used have taken place in its arenas. Since federation in 1901, and especially since the 1970s, the contests have swelled to involve the constitutional, and sometimes fractious political, relationships between Commonwealth and State governments.

The aim of this book is to display the complexity of the forest contests for, as in other sectors, the issues have to be understood more deeply than the media can depict before prospects for the future can be formed. The Introduction provides a brief description of the forests and their functions, the 'resource regimes' by which they are used, and the foundations of environmentalism which now challenge the older ideas of conservation and development. The first part of the book then provides a brief historical outline of how the forests were fashioned by economic and political forces before the 1970s, and how the ideas of forest conservancy, conservation, protection and forest development gained hold. The second part deals with three great changes of the 1970s and 1980s: the start of the woodchip export trade, the restructuring of the wood industries, and the ascent of environmentalism as a social and political force. The third part describes the numerous contests which resulted and the various, often unsuccessful, attempts of governments to resolve them. The final part looks at the prospects for the sector and considers how Australians might form its future.

A historical perspective is necessary because the forests of today are the accumulated result of past human activities and events. Many forests express a long and varied history of logging. In some, there are still trees older than white settlement whose rings and fire scars buried deep within their wood record their history. The forces and influences which fashioned them also created the reality which individuals strove to maintain or change. Some, more personally successful than the many, appear in this book as the leaders of their time. But the written historical record is selective; there is little in it of the countless timber workers whose often isolated lives were shaped by the same realities, and whose work is recorded only by its fading marks.

To elaborate the complexity of the forest contests, this book must leave much out. Most seriously, little can be included of the forests themselves or the changes to their ecology. The book is also limited to the land which has remained as forest and the main contests over its use for timber production; the great transformation of one-third of Australia's forest land for agriculture and settlement, and the history of a host of products important in their time yet trivial today – fuel wood, sandalwood, skins, wattle bark, and many more – remain topics for another day. The focus is on the public forests, and much less attention is given to those in private hands.

The forest contests seem never-ending. Just after the manuscript of this book was completed in October 1994, a major controversy erupted over the licensing of woodchip exports which exposed the inadequacies of processes and people to resolve the issues. A brief discussion has been added in the final chapter. But as the future can be expected to become progressively more difficult for the forests, as for Australia generally, such issues will need to be re-examined repeatedly. Creating a sustainable future will be a long journey; it is hoped that this book helps us start.

ACKNOWLEDGEMENTS

I am most grateful to Professor Henry Nix and my colleagues in the Centre for Resource and Environmental Studies of the Australian National University for their encouragement of this study. Les Carron, Andrew Claridge, Bruce Davis, Steve Dovers, John Formby, Marie Jamieson, Suzanne Hoverman, Kit Loughlin, Bill McCormack, Paul Parker, Ben Richardson, Alistair Sarre, David Sonnenfeld, Jacquie Tracey, Ian Watson, and members of the Australian Forest History Society are among the many whose ideas and knowledge have aided my understanding. Many organisations mentioned in the text and tables kindly provided information, and Roger Green did so about the media. Annie Boutland, Kay Dixon, Jane Goddard, Gloria Jackson, Sarah Jennings, Andrew Major, Warren Mortlock, Ralph Ogden, Kate Ord and Brittany Parker collected many of the materials for this and earlier studies on which I have drawn. Jenny Kesteven prepared many of the graphs, Mike Maslen of the Australian Heritage Commission prepared the map of wilderness, Phil Tickle of the National Forest Inventory provided maps of the forests which Kevin Lyon drew, and Tim Bonyhady advised on the selection of the cover. I am indebted to the library staff of the Australian National University for their courteous assistance, and to the Forest Products Association of New South Wales for allowing access to their press clippings files.

Heather Aslin, Les Carron, Suzanne Hoverman, Deb Fokey, Kit Laughlin, David Lindenmayer, Ian Penna, Angus Pollock, David Sonnenfeld, Jacquie Tracey, Jeremy Whitham, and two anonymous readers generously commented on parts of the draft manuscript but are in no way responsible for my emphases or the errors which remain. I am conscious that there is much to be done in writing Australia's forest history and hope that this book will stimulate further work.

To Jill Lane, the most patient of publishers, and all who have helped me, I give my sincere thanks.

John Dargavel
Centre for Resource and Environmental Studies
The Australian National University
Canberra

ABBREVIATIONS

ABS	Australian Bureau of Statistics
ACF	Australian Conservation Foundation
ACTU	Australian Council of Trades Unions
AFDI	Australian Forest Development Institute
AGPS	Australian Government Publishing Service
AHC	Australian Heritage Commission
ALP	Australian Labor Party
ANM	Australian Newsprint Mills Pty Ltd, and Ltd
ANPWS	Australian National Parks and Wildlife Service
ANU	The Australian National University
ANZAAS	Australian and New Zealand Association for the Advancement of Science
APPITA	Technical Association of the Australian and New Zealand Pulp and Paper Industry
APM	Australian Paper Manufacturers Ltd, later Amcor Ltd
APPM	Associated Pulp and Paper Manufacturers Ltd
AUSLIG	Australian Land and Surveying Information Group
AusTIS	Australian Timber Industry Stabilisation Conference
CALM	WA Department of Conservation and Land Management
CBCS	Commonwealth Bureau of Census and Statistics
CCV	Conservation Council of Victoria, later Environment Victoria
CRES	Centre for Resource and Environmental Studies, ANU

CROPS	Concerned Residents Opposing Pulp mill Siting (Tas.)
CSIR/CSIRO	Council for Scientific and Industrial Research / Commonwealth Scientific and Industrial Research Organisation
CSR	Colonial Sugar Refiners Ltd
ECF	Elemental chlorine-free (pulp-bleaching processes)
EGC	East Gippsland Coalition (Vic.)
EIA	Environmental Impact Assessment
EIS	Environmental Impact Statement
ERIN	Environmental Resources Information Network
ESD	Ecologically Sustainable Development
FAFPIC	Forestry and Forest Products Industry Council
FAN	Forest Action Network (Tas.)
FICA	Forest Industries Campaign Association
FIRM	Forest Industry Resource Management Group
GIS	Geographic Information System
IFA	Institute of Foresters of Australia
IUCN	International Union for the Conservation of Nature and Natural Resources, later World Conservation Union
LCC	Land Conservation Council (Vic.)
MOU	Memorandum of Understanding
NAFI	National Association of Forest Industries
NBH	North Broken Hill Ltd, North Broken Hill Peko Ltd
NEFA	North-East Forest Alliance (NSW)
NFAC	Native Forest Action Council (Vic.)
NSSC	Neutral sulphite semi-chemical pulping process
PMP	Preferred Managament Priority system (NSW)
RAC	Resource Assessment Commission
SEFA	South-East Forests Alliance (NSW)
SSDA	Social Science Data Archives, ANU
TCF	Totally chlorine-free (pulp-bleaching processes)
TPFH	Tasmanian Pulp and Forest Holdings Pty Ltd
TWS	Tasmanian Wilderness Society, The Wilderness Society
VAUS	Value Adding Utilisation Scheme (Vic.)
VNPA	Victorian National Parks Association

PATTERNS OF AUSTRALIA'S FORESTS

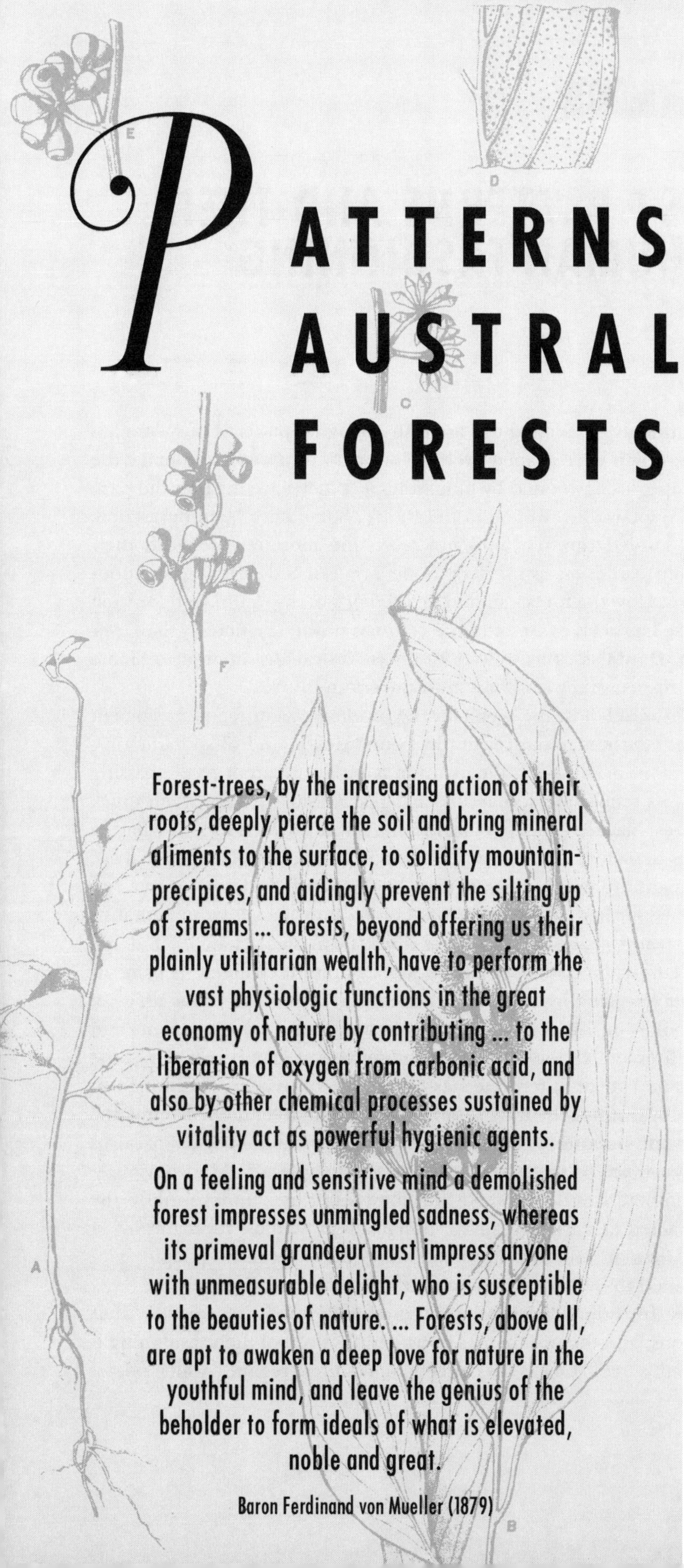

Forest-trees, by the increasing action of their roots, deeply pierce the soil and bring mineral aliments to the surface, to solidify mountain-precipices, and aidingly prevent the silting up of streams ... forests, beyond offering us their plainly utilitarian wealth, have to perform the vast physiologic functions in the great economy of nature by contributing ... to the liberation of oxygen from carbonic acid, and also by other chemical processes sustained by vitality act as powerful hygienic agents.

On a feeling and sensitive mind a demolished forest impresses unmingled sadness, whereas its primeval grandeur must impress anyone with unmeasurable delight, who is susceptible to the beauties of nature. ... Forests, above all, are apt to awaken a deep love for nature in the youthful mind, and leave the genius of the beholder to form ideals of what is elevated, noble and great.

Baron Ferdinand von Mueller (1879)

Mueller, the great colonial botanist, studied the forests as a scientist, disseminated information to the public about their timbers, oils and barks, and revered them as the Creator's work; he complained of waste, sought to reduce timber imports, advocated plantations, contributed his ideas for better forest management to government inquiries, and insisted on 'the stern necessity of securing a due relation of forests to territory, of woods to climate, of timber to industries'. All these themes run through this book and remain familiar today.

INTRODUCTION

FOREST PATTERNS AND THEIR HUMAN FASHIONING

The infinitely detailed and ceaselessly changing patterns of Australia's forests, which were shaped over evolutionary time to fit and make the landscape, are fashioned by human hands; they have been in the past, they are today, they will be in the future. Some have been devastated, others tended carefully, and now even the most remote need their remaining wildness protected; none are isolated from disputatious humans. How the forests should be used or not, by whom, and for what purpose has been contested for a century; now more fiercely than ever before. Doubtless, this will continue as Australians attempt to create what prospects they can for a most uncertain future.

The forest landscape is equally the product of nature, of the ancient cycle of reproduction, growth, decay and death, and of sporadic fires and cyclones. This natural dynamism is rarely apparent to the passing viewer; its scenes are serene, its changes are slow, except when quick bushfires blacken it or in northern Australia when cylones flatten swathes across it. By contrast, the human dynamism obtrudes sharply; the loggers are noisily at work, trees on familiar hillsides disappear, raw roads are pushed where none was before, and now environmentalists protest their claims to 'save' the forest from such changes. But the forest landscape evidences past as well as present changes; its patterns are detailed stand by stand, tree by tree – or 'fashioned' in the usage of this book – in distinct stages by tracks and tramways, settlements and roads, fires and cyclones, clearing and fellings at many times. Thus, the landscape can be interpreted as a layered construction fashioned in stages by intimately interwoven social, spatial and environmental factors and by natural changes to its patterns.[1] But the natural patterns and dynamism of the forest acted on the people too. Most obviously, these imperatives determined how they could use it; more subtly, the forest acted as part of their Australian identity; most dramatically, in forest fires it even killed them; and more recently, people have responded to the moral imperative to preserve its uniqueness.[2]

This Introduction gives a brief overview of the forests, their functions, and the two themes by which the story of their fashioning is told in this book. One is the concept of 'resource regimes' which relates the changing face of capitalism – the engine of Australian development – to the industrial use of the forests; the other is the through the evolution of ideas of conservation and preservation which, in attempting to modify or halt industrial use, infuse past and present contests.

Natural patterns

The interpretation in this book starts with the definition of just what sort of woody vegetation comprises a 'forest'. This is not obvious in Australia, where the closed, dense 'forests' of the high-rainfall regions fade gradually to the sparse, open 'woodlands' of much of the dry interior. This book is concerned with contests over the use of forests as timber resources, so the conventional Australian definition of forests as having the capacity to produce timber – sawlogs, pulplogs, poles, piles, railway sleepers and the like – is quite suitable. Thus, even in its definition, a forest is a both natural and human construct.

In the world pattern, Australia's forests, their plants and creatures, are distinct, their evolution having taken its own path. Biogeographers account for this by tracing the drift of continents from a mythically-named Gondwanaland, the surge and retreat of ice ages, the rise and fall of seas, and the chill and returning warmth of climates.[3] Among the timber trees, they find links in many groups of similar species, or 'genera', across the southern continents from their ancient touching: the *Araucaria* (Australia's hoop and bunya pines) and *Nothofagus* (myrtle beech, for example) can be found in South America. But the great genus *Eucalyptus*, which so conspicuously characterises the landscape, is almost entirely Australia's own. Of its 600 species, only a handful occur in the islands to the north, Papua New Guinea and Timor, and only one occurs as far away as the Philippines. Somehow – there are different theories – much of the Australian flora, including the eucalypts, evolved with hard, thick-skinned leaves and other 'scleromorphic' features which enabled it to survive Australia's poor soils, droughts and fires.

The relatively small areas of tropical and sub-tropical rainforests with their multitudinous species – which curiously coexist with wet sclerophyll forests in places – can be related to the flora of Papua New Guinea, Melanesia and South-East Asia through many common genera. However, Australia has been isolated long enough for most individual species in each genera to have become unique to Australia, or 'endemic'. The animals are singularly so. They are mostly marsupials, of which there are few outside Australia, while the egg-laying mammals

– platypuses and echidnas – are unique. About one-third of the bird species are endemic.

Nature endowed Australian forests richly with unique species, but poorly in area. In the current climatic period the forests have never covered more than about 9 per cent of the land; their extent and composition are determined by those great environmental variables – climate, topography and soil. Needing at least 600 millimetres of rain a year, they could grow only in the belt which reached from the east and south-east coasts somewhat over the Great Dividing Range and, for cypress pine, stretched slightly further out into drier country in New South Wales and Queensland. There are also forests in the south-west of Western Australia, and in much of Tasmania. The forests reach up to all but the highest mountains and in the tropics stretch with a great belt of paperbark down to the seasonally swampy areas and out into mangroves.

The great continental pattern is made of many types of forest, most dominated by the ubiquitous eucalypts or intermingled with them (Table 0.1). For example, many of the rainforests, especially the temperate ones in Tasmania, have eucalypts emerging here and there to tower over the smaller trees that characterise the type. The relative extent and tenure of the 43 million hectares now remaining is shown in Figure 0.1.

Within the major types of forest, smaller patterns are made – catchment by catchment, patch by patch – by variations in topography, climate, geology, soil and water. Forests on north-facing slopes and ridges, for example, are generally less diverse and grow more slowly than those in gullies or on more sheltered slopes. Cliffs, rocky outcrops, lakes and swamps break the forest cover. Deep, young, fertile soils on which trees grow best are not plentiful; more general are the old, often shallow soils in which only hardy species can grow, and that slowly. These smaller patterns become more detailed still, because most of the forests have a mixture of tree species: typically a dozen or so in a few hectares of eucalypt forest, or a hundred or so in the tropical and semi-tropical rainforests. Exceptions which grow in almost pure stands are the magnificent south-east mountain and alpine ash forests, the south-west jarrah and karri forests, the river red gum forests along the Murray River and the cypress pine forests.

Each of these types of forest, captured in the human eye by its commanding trees, has its set of plants and animals bound together by the cycles of water and nutrients, oxygen and carbon. Some species are unique to a particular forest type, as Leadbeater's possum is to mountain ash forests for example, while others can be found in several. Similarly, the degree of inter-dependence varies; some species are

Table 0.1 Major types of forest

Type	*Region*	*Main species*	*Area (million ha)*
Eucalypt forest			
South-west dry	South-west of WA	Jarrah-marri	2.2
South-west wet	South-east of WA, high rainfall areas	Karri	0.2
South-east dry	NSW, Vic. and south-east of SA	Red stringybark, yellow and white box	6.4
South-east wet	High rainfall, montane, NSW, Vic., Tas.	Blue gum, messmate, and manna gum	4.1
South-east ash	Most productive part of south-east, wet eucalypt forest	Mountain ash, alpine ash, and shining gum	1.5
South-east coastal	Coastal lowlands, NSW, Vic.	White stringybark and silvertop are widespread in south; apple and red bloodwood and some rainforest species occur in north	3.8
Central coastal	Widespread in NSW, QLD	Many species, including flooded gum, tallow wood, blackbutt and brush box	4.2
North-east coastal	From northern NSW to Cape York	Many species, including pink bloodwood, carabeen, forest red gum	2.7
River red gum	Widespread with extensive stands along River Murray	River red gum	0.5
Rainforest			
Tropical	Queensland	A great many; hoop pine is widespread	1.1
Sub-tropical	South-east Qld and northern NSW	Many, including yellow carabeen and blush tulip oak; eucalypts are sometimes present	0.4
Temperate	Cool, wet Tas., Vic., south-east NSW	Myrtle beech; eucalypts can be present	1.0
Mixed forest and woodland			
Native pine	Inland NSW, Qld	Cypress pine	4.0
Northern	Northern Australia	Darwin stringybark, lancewood	10.1
Mangrove and swamp forest	Mostly on flood plains of northern Australia	Paperbark and eucalypt swamp trees	0.9
Total			43.2

Source: Adapted from Resource Assessment Commission (1992), vol. 1, box 3.1.

dependent entirely on others for food, shelter or support, others less so. Ecologists see each type of forest, or 'ecosystem', as a functioning whole made of all its inter-connected elements. Their notion of the forests' responses to change is a dynamic one. These responses do not always take the same path after events such as cyclones, bushfires, or felling, so that the mix of species may alter somewhat, which makes the changing pattern even more intricate. Few of Australia's forests are as

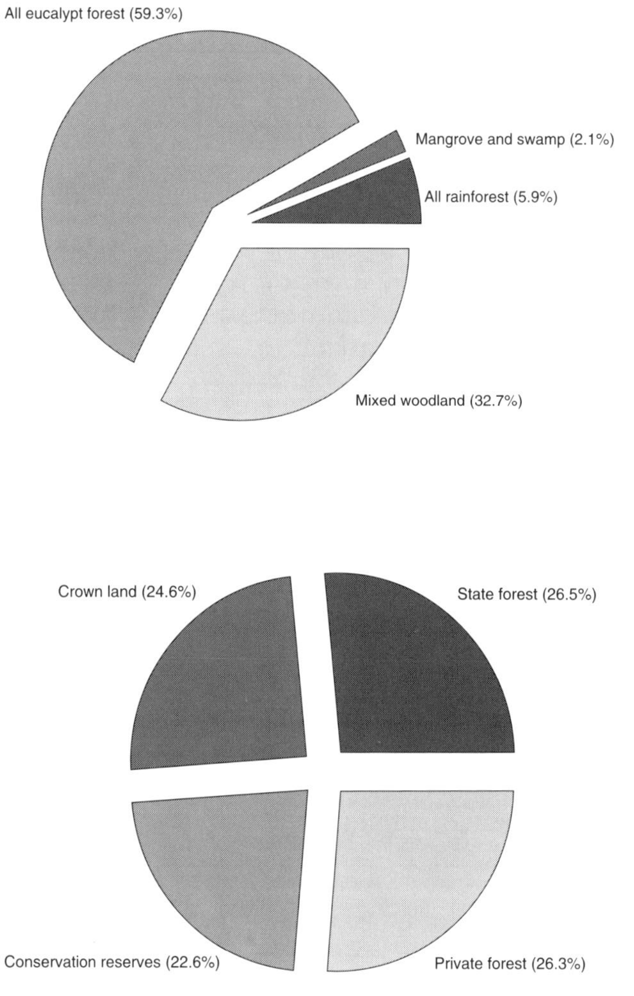

Figure 0.1 Extent and tenure of major forest groups
Note: Total area 43.1 million hectares.
Source: Resource Assessment Commission (1992).

simple as the great conifer forests of the northern hemisphere, and only the northern rainforests are as complex as those in the tropics.

Present forest landscapes

The present forests are those left, or saved, from two centuries of deforestation. Wherever climate, soils, topography and access were suitable, the forests were cleared for agriculture and settlement, although, as the case for forest conservation mounted, some were kept as state forests and other reserves. By the 1980s, about 26 million

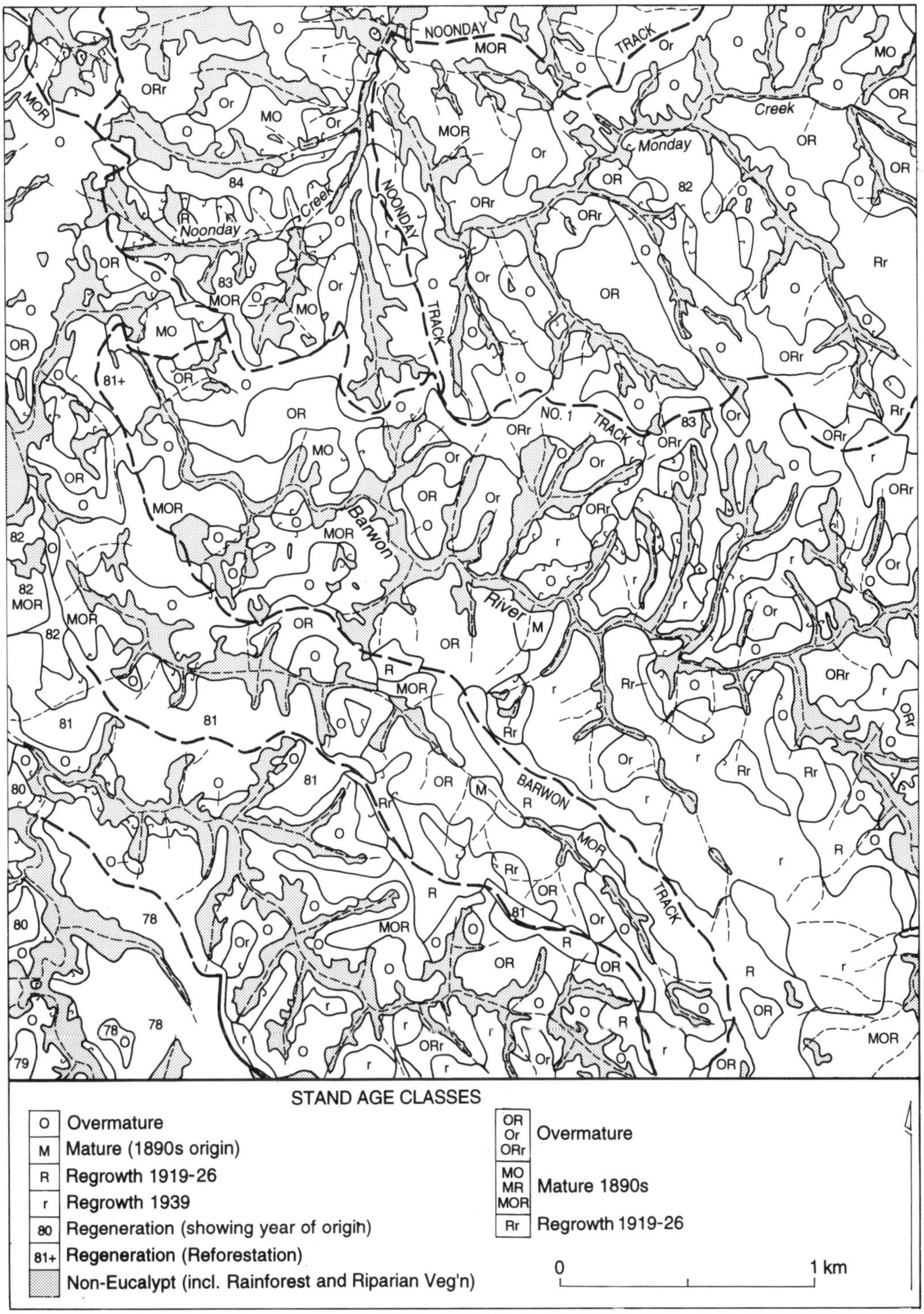

Figure 0.2 Intricate pattern in mountain ash forests

Source: Redrawn from a portion of a stand type map prepared by the Resource Assessment Section, Department of Conservation, Forests and Lands, Victoria, 1985.

hectares had been deforested, leaving 43 million hectares as the forest estate (Figure 0.3) which has now been supplemented by 1 million hectares of forest plantations.

The settlers cleared forests on the better soils and, with exceptions, the flatter land on the rough principle that the tallest trees indicated the best soil. Some types, such as the lowland tropical rainforests which were cleared for growing sugar cane, were so extensively deforested that little remains. Generally, the forests remaining today are on poorer, steeper, or less accessible land. Even in the most forested regions, the forests occupy the hilly background in settled landscapes; in few districts are they the most important land use.

The forests provide important ecological, economic, social and cultural functions and values in their landscapes (Table 0.2). Many are public or intangible benefits, others marketable commodities; some complement others, or can be managed compatibly with them, while others conflict either inherently or because some particular group uses them. In this complexity of functions and values lies the heart of the contests over how they are used and who has the right to do so. The poor natural endowment of forests, their wide dispersion around the continent, their mixed and variable pattern, and their subsequent reduction and fragmentation by settlement constrained the development of Australia's wood industries. Indeed, Australia has long been a net importer of forest products. It is not because the industries are large in an international sense that their impact is so contested, but because the resource is small and its character unique.

Resource regimes

The right to use the forests, like other natural resources, can be considered legally in terms of private property, state property and common property, each of which confers different legally enforceable, rights and responsibilities. These are essentially sets of social and political relationships.[4] They are 'human creations whose purpose is to manage people in their use of environmental resources'.[5] The structure of rights and duties which characterises the relationship between individuals, or between individuals and the state, varies between resources and is known as a 'resource management regime', or 'resource regime'. The nature of the regime determines how the resources are used, and hence how they are spatially and environmentally fashioned.

As three-quarters of Australia's forest lands are in the hands of the state, state property and its management are the central to this book. It

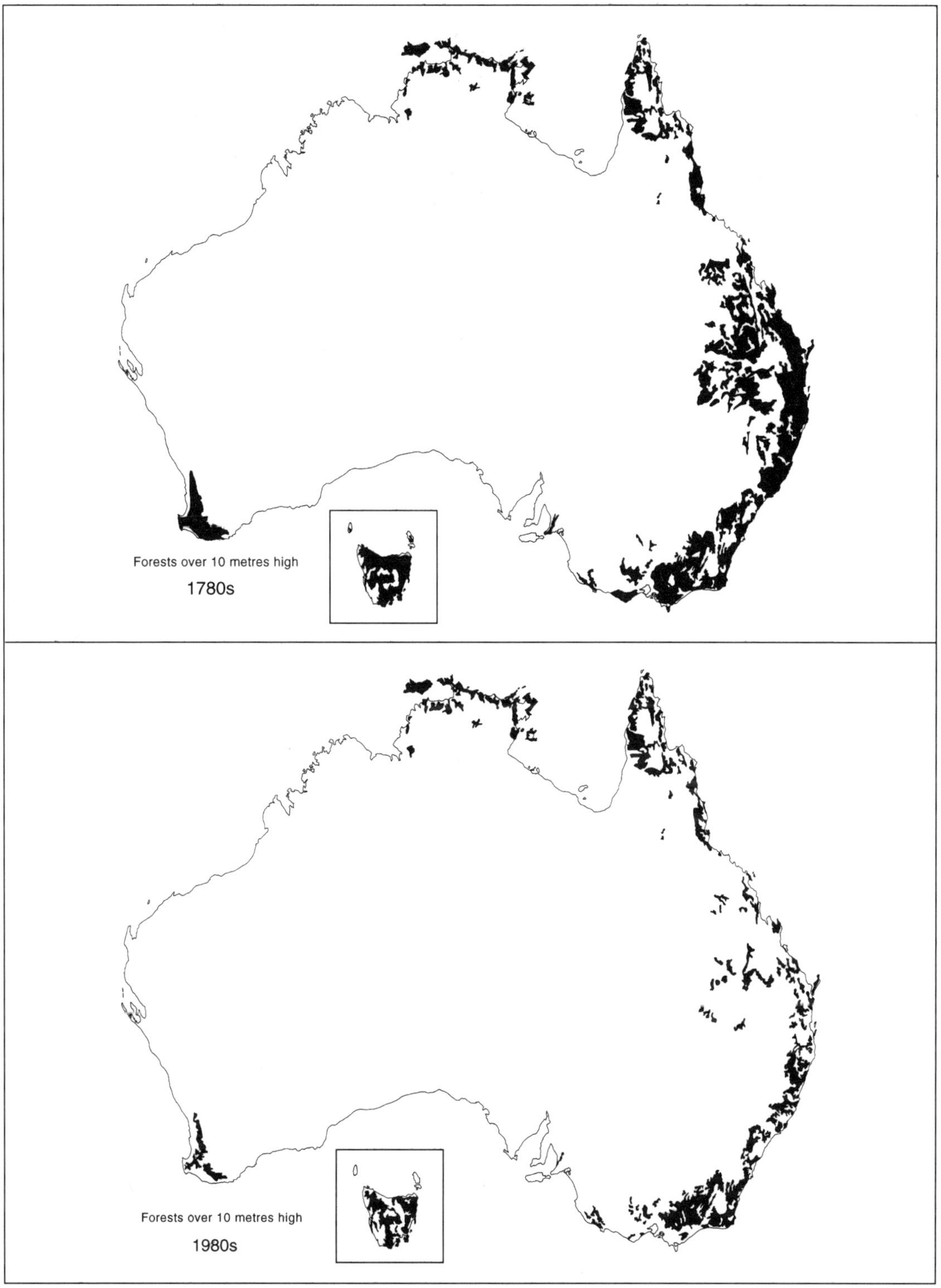

Figure 0.3 Forests and woodlands in the 1780s and 1980s
Source: Drafted by K. Cowan from AUSLIG, *Atlas of Australian natural resources* (Canberra: AGPS, 1990), vol. 6, *Vegetation*.

Table 0.2 Forest functions

Category	*Function*
Ecological	
Climatic	Prevent wind erosion and provide shelter for stock
	Sequester carbon
Hydrology	Meliorate run-off and hence floods and drought
Soil	Protect against erosion
	Maintain fertility
Fauna and flora	Provide habitat
	Maintain biodiversity including rare and endangered species
Economic	
Commodities	Industrial wood – sawlogs, plylogs, pulpwood, sleepers, piles, etc.
	Fuelwood
	Grazing
	Minor products – bark, flowers, medicines, food, etc.
Public goods	Water
	Game animals
Industrial research	
Social	
Employment	Protection and management
	Producing goods and services
Decentralisation	Support small rural communities
Recreation	Driving, picnics, walking, horse-riding, camping, trail-bike riding, etc.
Noise reduction	
Environmental education and research	
Cultural	
Rural communities	Fabric of life, social community
Heritage	Aboriginal, archeological, natural, industrial
Artistic	Aesthetics, landscape
Aboriginal	Spiritual identity
Spiritual	Religious, spiritual refreshment
Wilderness	Experience, existence value

is necessary to look beyond the concept of property, which is defined by how far the rights to use the resource are devolved into private hands or retained by the state, and consider other relationships that affect the users and the forests. Most important are the administrative procedures for regulation and management, and the policing and legal procedures for obtaining compliance.[6] These relationships – expressed in a mass of licences, leases, regulations and administrative customs – do not arise in a political vacuum; rather they are the consequence of contests between various users and interests. The state appears, in this book, as an arena in which the contests occur.

The concept of a general resource regime needs to be refined to specific regimes, because the state conceded very different resource rights to particular stages and types of industrial capital. Much of the story of the forests concerns the disputes and accords between large and small, foreign and local capital. The concept of specific forest resource regimes used in this book includes the (intra- and inter-class) relationships influencing the state's behaviour; the resource rights allocated; the typical organisational form of industry; and the ideas, beliefs and practices which supported it. Typical rights, responsibilities and relationships involved in each regime are depicted in Table 0.3, although the details varied considerably from State to State and changed with time.

Table 0.3 Major forest resource regimes

			Typical resource rights over public forest land				
Regime	*Capital*	*Labour*	*Products*	*Form*	*Extent (hectares)*	*Duration (years)*	*Quantity (m^3/year)*
Small sawmills	Competitive: family firm, private company	Owner, family, wage	Sawlogs	Exclusive licence	100 to 1000	1–10	1000 to 10 000
Large mills (sawmills, board mills)	Competitive: private or public company	Wage	Sawlogs, now some pulpwood	Special licence, agreement	20 000 to 200 000	15–42	50 000
Exclusive concession (pulp mills)	Monopoly: public company	Wage	Pulpwood some sawlogs	Legislated agreement	Up to 500 000	88	1 000 000
Integrated concession (pulp mills, woodchips)	Global monopoly: public company multi-national corporation	Wage	Pulpwood some sawlogs	Legislated agreement	Up to 1 000 000	15–88	1 500 000
Logging contractors which supply all regimes							
	Contract, sub-contract: family firm	Owner, family, piece-wages	No resource rights in public forests				

Ideas and ideals for conservation and preservation

Ideas and ideals about how the forests should be used helped to form and change the various regimes by expressing or countering the material interests. Most ideas have long histories, and the ideals or visions taken up by groups of people as causes in the political process evolved in stages, often in reaction to the consequences of forest use; the environmental movement is the largest and most recent example. Older ideals persisted, so that the landscape of ideas, ideals and causes became a layered social construction like the forest itself. It too has complex patterns of cultural traditions, science, art, philosophy, ethics,

Table 0.4 Evolution of ideals for Australian environmental management

Period	*Characteristics*
Aboriginal management	*Subsistence:* religious identification with the land
Exploitative pioneering: nineteenth century onwards	*Exploitation:* basic survival regardless of the environmental consequences, followed by development of export staples – whale oil, cedar, wheat, wool, meat, etc. *Colonialism:* Australian development geared to British needs and culture *Enlightenment thought:* progress, growth, development, rationality, the advance of colonial science and increase of productivity through applications *Yeoman farmer/agrarianism:* ideals of settlement dominated land policy
National development and 'wise use' of resources: 1880s onwards	*Imperialism:* Australian development geared to Empire needs *Development:* state promotes industrial and rural development through railway, irrigation, electricity and other schemes *Employment:* state relieves unemployment in 1930s and promotes industrialisation to create full employment in 1950s and 1960s *Conservation:* state intervention for 'wise use' of resources
Modern environmentalism: 1960s onwards	*World-wide wave of social change:* environmentalism arises as one of several new social movements with many strands: • Moderate environmentalism and conservation as 'wise use' which believed reform within existing Western humanist values and institutions should achieve development in harmony with nature. • Radical environmentalism which believed that non-human life has inherent or intrinsic value independent of human use. It characterised its 'deep ecology' position as 'ecocentric', and the moderate position as 'shallow' and 'anthropocentric'. *Preservation:* the right of species to exist irrespective of human society becomes accepted *Wilderness:* areas are excluded from development *Ecologically sustainable development:* adopted as a goal to maintain biological diversity and ecological processes, while also securing development

Source: Adapted from Frawley 1994.

morals and religious principles which have been examined in many ways beyond the scope of this book. Most important for the recent forest contests is the emergence in modern environmentalism of an 'ecocentric' philosophy, which places an inherent value on the existence of non-human life irrespective of its human use.[7] The stages and their key elements can be displayed in a simple overview (Table 0.4). The challenge for the future is to devise and install resource regimes which will enable Australia's forests to fulfil their natural and social functions and meet the evolving ideals of society.

FASHIONING

1850s TO 1960s

Our first-class Eucalypts are few in number, the areas upon which they grow are limited, and the supply from such areas has in the past been so recklessly drawn upon by splitters, sawmillers and timber getters that at the present time, as far as I can see, we can do no more than barely supply our own wants. I see no sufficient margin at the present, or even in the near future, for any extensive export of first-class timbers. Indeed our future supplies are questionable unless steps be taken to protect our remaining gum forests, and to encourage the growth of suitable young trees.

Alfred Howitt (1895)

Forestry will not flourish until it is recognised that with forests, as with other things, there is a limit to the process of getting something for nothing.

Sir Roy Robinson,
Chairman of the British Forestry Commission (1922)

So did the famous explorer assess Victoria's forests after less than half a century of appreciable exploitation. The next two chapters describe the resource regimes which fashioned Australia's forests until 1970, primarily to meet the needs of the rapidly growing country, but also for a large timber export trade mainly from Western Australia. Sawmilling and the new pulp and paper industry placed rapidly increasing demands on Australian forests during the economic boom of the 1950s and 1960s, so that new regimes had to be devised.

Paralleling the story of how the forests were used is another, told in Chapter 3, of the gradual rise of conservation to take the very steps that Howitt, Robinson and others advocated. Together, these three chapters set the stage and institutions for the contests which erupted from the 1970s.

CHAPTER ONE

FORESTS FOR HOME AND EMPIRE

The Australian landscapes discovered by European explorers as a new world are still haunted by their ancient past. The woodchips now being exported to Japan come from forests first fashioned by Australia's original people and then refashioned by the successive regimes of capitalism. This chapter looks at how the new society used the forests in its new land. The first part sketches the period before 1850 as a prelude to the second part which describes the rise of the modern timber industry and its energetic exploitation of the forests for home consumption during Australia's long economic boom from the gold rush to the 1880s. The third part describes how new regimes were applied to meet the needs of the expanding British Empire during the period up to World War I.

Manual prelude

Australia had long been an isolated continent when the original people arrived 40 000 or 50 000 years ago to find a largely unfamiliar flora and fauna. The climate warmed and cooled and warmed again, shifting the position of the coasts, and the boundaries and composition of the forests.[1] The people spread across the continent, diversified, separated by distance, territorial custom and often mutual hostility, and came to identify themselves with their land in profoundly spiritual ways.[2] Although their use of the forests varied between the groups and types of forest, fire was the ubiquitous but capricious tool which at times could escape even the most careful controllers.[3]

Aboriginal firing did much to fashion an open grassy woodland landscape which was the most productive habitat for the grazing and arboreal animals that they hunted. For example, in the Central Plateau of Tasmania, firing maintained a landscape of eucalypts and grass against the possible invasion of rainforest species.[4] In other places, it pushed back the margins of rainforests, and may have greatly reduced the area of monsoon rainforest in tropical Australia, although otherwise forests were probably little affected by the harvesting of their fruits and animals.[5] Overall, it seems clear that the long period of Aboriginal use and management made far lighter marks on the forest landscapes than those of the subsequent invaders.[6]

In 1642 the Dutch East India Company sent Abel Tasman in search of spices, silks and precious metals to 'discover, the partly known and still unreached South and Eastern land [Australia)] ... for the improvement and increase of the Company's general welfare'.[7] He found Australia singularly unimpressive and sailed on, leaving little but names for Van Diemen's Land (Tasmania from 1856) and a few other places on his careful charts. About the same time fishermen from Macassar started to visit the north coast of Australia to collect and process trepang for the trade with China where it was prized as a delicacy.[8] They used the mangroves and took some she-oak back – probably Australia's earliest timber export – but their overall impact was trivial.

It was not until the late eighteenth century that Australia began to be absorbed into the European world system, and by that time it was a different world. Britain had passed through its political revolution and by the 1780s was starting the industrial one which was to make it the 'workshop of the world'.[9] At Waterloo (1815) it finally emerged as the hegemonic power. The tale of how immense social upheaval, endemic poverty, a savage legal code, strategic and trade concerns led to the establishment of a penal colony in remote Australia has been told elsewhere.[10] Here it can be noted that Australia was also incorporated scientifically, in ways epitomised on Cook's first voyage to the southern oceans and Australia (1768–71) by all the specimens of plants and animals which the worthy Joseph Banks collected. It was back in London that they were carefully drawn and classified; it was there that the new scientific understanding of Australia's forests commenced.

The fleet full of hapless convicts and their jailers which landed in Botany Bay in 1788 needed to deal with their startling new environment in a most immediate way. The main task was to support the settlement with food,[11] but timber was needed too for houses, barracks and huts to shelter the resplendent Governor, his officials, officers and troops and eventually the convicts; and it was the convicts who had to cut it. They did so in a sawpit at Botany Bay even in the few days of January 1788 before the First Fleet moved to its permanent landing at Port Jackson. Suitable pit-sawing sites were then found along the arms of Port Jackson and the banks of the Paramatta River, whence the timber was shipped across to Sydney. Soon after Collins landed in Van Diemen's Land in 1804, timber was cut at the King's Pits, on the lower slopes of Mount Wellington, and a convict gang dragged a rough carriage of timber to build Hobart Town.

Wood was vital to the emerging colonial economy. It walled huts and houses, and slabs of bark or wooden shingles roofed them. Wooden palings fenced town gardens, stout posts and wooden rails made the stockyards, while crudely piled logs made the early fences. Wood made the ships, carts and carriages; it made water pipes and parts of machines; it made butter boxes and tight barrels for salt meat and whale oil. When bricks were made, wood fired the kilns as it did the bread ovens. Coal was brought to Sydney from the convict station on the Hunter, but apart from this and whale oil for lighting, wood was the colonial fuel. When steam engines came, it fired their boilers. By trial and error, the best sort of trees were found: she-oak for hut poles and shingle roofs, gum and box for sound timber, ironbark and stringybark for split slabs, posts and rails.[12] In 1790 the soft and easily worked red cedar that was good enough for joinery was discovered along the Hawkesbury River, and in 1815 the superb boat-building timber, huon pine, was found in the south-west of Van Diemen's Land.

Timber was sawn, split, and hewn with immense manual effort. Once cut, the flitches of sawn timber and the bundles of shingles and palings had to be carried, most often on human backs, out of the forests to wherever they could be loaded on to carts or ketches. Luckily for the cutters, cedar and huon pine grew near water and, unlike most Australian woods, their logs floated and could be made into rafts for towing at least part of the way. Production was organised in three regimes: prison farms, petty production and merchants.

Prison farm regime

Timber was produced by the convicts first as a state activity and later as a punishment. State production faced dilemmas: for efficiency, the convicts had to be scattered through the bush so that they could cut timber as close to the stump as possible; but to prevent them escaping, they had to be concentrated. In Van Diemen's Land, for example, once the resources at the King's Pits had been exhausted in 1819, new sawing stations were established in the then remote Northwest and Birch's Bays, south of Hobart. At first, large solid sawpits were constructed and the logs dragged or rolled to them by hand. The convicts were easily controlled, but the trees nearby were soon cut out. The central sawpits were replaced by temporary, lightly-constructed pits scattered through the forests, reducing the effort required to move logs long distances by hand. The resulting supervision problem was solved by setting the sawyers a weekly task to complete, after which they could work 'on their own account', being paid for what they produced.[13]

But the British government demanded that terror in the penal colonies be sufficient to quieten the populace at home. Transportation was not enough without the lash, and the lash was not enough without the secondary punishment stations at Norfolk Island, Newcastle, Maria Island, Port Arthur, Moreton Bay and Macquarie Harbour, where 'man lost the aspect, and the heart of man'.[14] Timber production was part of their barbarous regimens, in which 'caterpillars' of convicts were forced to carry the logs on their shoulders. At Macquarie Harbour they rolled huon pine logs great distances out of the bush and formed them into rafts. 'In this service, life was sometimes lost; and the miserable workmen, diseased and weakened by hunger ... often passed hours in the water.'[15] Many chose death in the bush rather than endure it.

Petty production and merchant regimes

The convict stations produced wood mostly for public works, so that commercial supplies were needed for the growing towns and cities. Manual production required few tools and so suited a largely ex-convict workforce which cut, sawed, hewed, split and lugged timber well into the 1870s. Many of the cutters and splitters were part of a largely itinerant rural workforce which worked variously as petty producers, contractors to landowners and farmers, or as piece-workers for timber merchants.

Sawyers' camp in the Shoalhaven district, NSW.
Sydney Mail, 29 December 1883. (National Library of Australia)

The colonial state introduced general controls over the entire workforce through Masters and Servants Acts, and specific controls over the timber workers through licensing. Employers had been required to obtain permits to fell cedar since 1802, but from 1828 in Van Diemen's Land and from 1838 in New South Wales, cutters had to obtain a licence from their local magistrate, provided that they were eligible 'in point of character'. As the Executive Council said in Van Diemen's Land, licensing 'is intended to be a matter of Police rather than Revenue'.[16]

While independent petty producers could meet many local needs, they did not have the capital or organisation to transport timber long distances and sell it in distant city markets, nor did many workers save enough to provision themselves for long periods of camping in the forests. Some colonial merchants recruited gangs of cutters, transported them to cutting sites and supplied them in the notorious 'truck' system, providing food on credit against the value of the timber cut.[17] By the 1840s specialist timber merchants began to establish themselves in the growing cities.[18] Their timber was largely shipped along the coasts or across Bass Strait, because land transport was so expensive.[19] Small quantities of cedar were also exported and some Baltic deals and battens imported, mostly from Britain. Notably, little of the production was mechanised. Only about twenty small sawmills were built in prior to 1850; a few of these were driven by horses, several by water wheels, and about eight by steam engines; several seem to have been short-lived or operated only intermittently.[20]

Mechanising the forest industries

The modern timber industry started during the third quarter of the nineteenth century, the 'golden era of capitalist growth' when gold and wool stimulated a long period of self-sustaining growth. Moreover, new colonies had been started without the convict 'stain' (the Swan River Colony in 1829, the Port Phillip District in 1835, and South Australia in 1836);[21] and high shipping costs, humane voices, and colonial opposition stopped Britain transporting convicts to New South Wales in 1842, and to Van Diemen's Land in 1852. The flourishing colonies came to govern themselves; Swan River became Western Australia, Port Phillip became Victoria, the dreaded Van Diemen's Land became Tasmania, and Queensland separated from New South Wales. Australia gained an attractive image with a healthy climate prescribed for those who could afford it, such as Andrew Wilson who started the Maryborough timber firm of Wilson Hart in 1863.[22]

It was the gold discovered in Victoria in 1851 that rushed immigrants from around the world. In ten years almost three-quarters of a million people were added to Australia's white population of 405 000. Notably there were more young women, so that natural increase came to double immigration; by 1880 there were 2.2 million people.[23] The colonial economies expanded and diversified, manufacturing increased, banks and great mercantile, pastoral, and mining companies prospered, while a rising urban bourgeoisie and new settlers demanded land and a political voice. The rapidly increasing population had a prodigious appetite for timber. Not only were more houses needed for the flood of people, but the poor gradually moved into houses, while the rich built elaborately fitted mansions. The existing manual timber industry could supply neither the quantity nor the quality suddenly required; a new order was needed, and quickly. It had three components: imports of softwoods increased, sawmills were constructed to mechanise production, and some timber workers continued to split and hew manually in the forests.

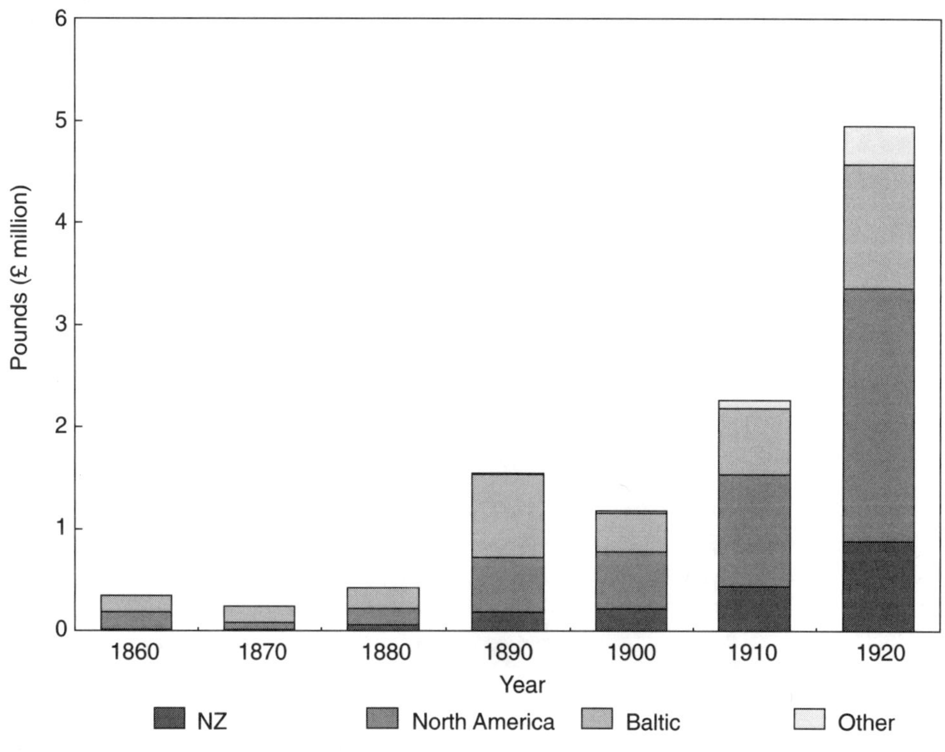

Figure 1.1 Imports of timber, 1860–1920
Source: Statistical Registers.

Trade in timber

Booming timber markets in Melbourne, Sydney and Adelaide attracted imports from neighbouring colonies and from other countries. The latter were almost all of softwoods, partly because of their natural shortage in Australia, and partly because they were available cheaply from the great producing regions of the northern hemisphere and from New Zealand.[24] New Zealand was also the destination of Tasmanian timber shipped to the Otago gold rush of the 1860s. Otherwise, small quantities of cedar and other special timbers were exported to Britain, and some timber was shipped or transhipped to various Pacific islands.

Rise of sawmilling

Mechanisation proceeded rapidly. Within four years of gold being discovered, twenty-five sawmills were built in Victoria and by 1860 there were sixty-four. Most were in the goldfields regions, but mills were also built in and near Melbourne and Geelong. Tasmania, with good timber close to sheltered harbours, built twenty sawmills in the 1850s, primarily to supply Melbourne and Adelaide. However, further expansion was

Timber workers.
(E.J. Brady, *Australia Unlimited*, 1918; National Library of Australia)

inhibited when Victoria introduced tariffs in 1871.[25] New South Wales was not so immediately affected by the gold rush, so that expansion there proceeded at a steadier rate. South Australia was limited by its small area of native forests: although there were twenty-six sawmills operating by 1870, most were resawing imported timber. Queensland, with a tiny domestic market, mostly produced logs of valuable species for sawing or veneering in southern mills.[26]

The booming demand for timber continued so that the number of sawmills in New South Wales and Victoria doubled between 1875 and 1890. However, the crash of Melbourne's building boom in the 1890s drove half Victoria's sawmills out of business. New South Wales was less affected, though it closed 150 of its sawmills, mostly very small ones. Queensland more than doubled the number of its sawmills during the 1880s and continued to expand to have 160 by the end of the century.

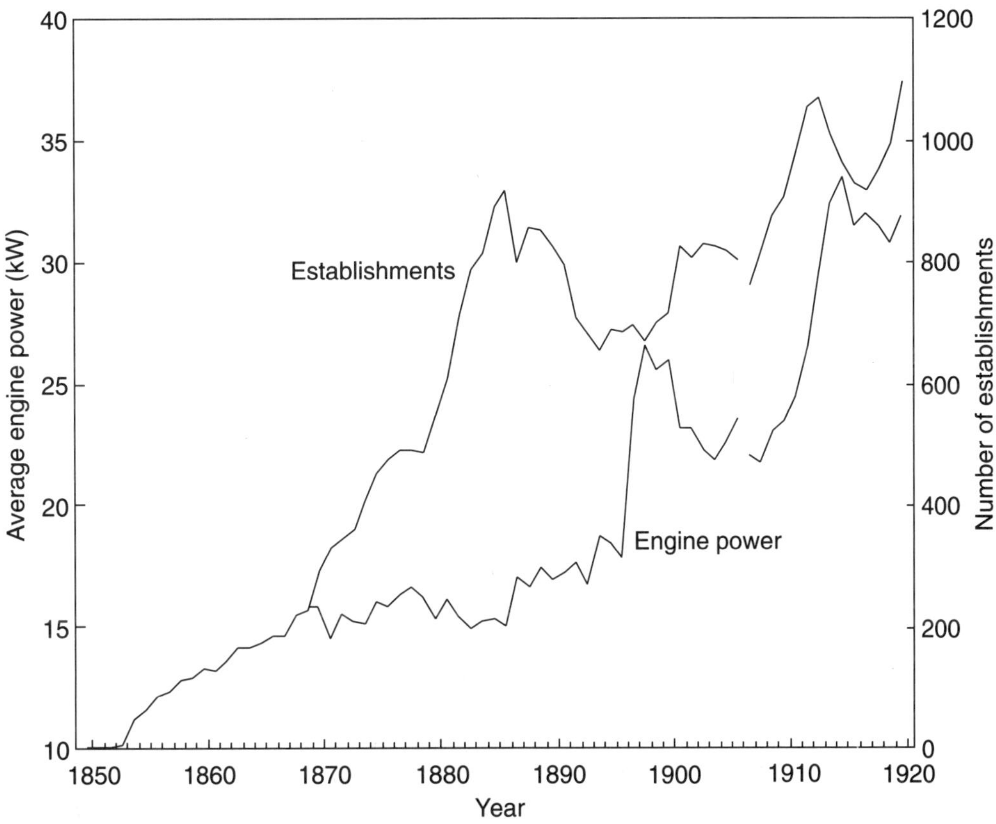

Figure 1.2 Sawmills, number and average engine power, 1850–1920
Note: Not all colonies reported engine power each year. The average was calculated from those reporting.
Source: Statistical Registers, 1850–1906; CBCS, *Secondary Industries Bulletin*, 1907–20.

Table 1.1 New South Wales sawmills, 1880

Power source		*Employees*	
	Number	*Total*	*Average*
Country mills			
Steam	182	1509	8
Water	21	98	5
Horse	5	10	2
City mills			
Steam	35	661	19

Source: NSW, *Statistical Register*, 1880.

By 1880, the structure of the sawmilling industry in the eastern colonies was clearly developed. There were 500 sawmills, mostly steam-driven. They operated as individual or family-owned concerns, although there were a few large, company-owned mills. The sawmills were differentiated into two main types, though their actual functions often overlapped. The forest mills cut logs directly from the forest and generally produced rough-sawn timber which was sold either 'green' as it was, or partly air-dried, to local customers or through city timber merchants. The town or city mills were larger affairs, often owned by timber merchants. They resawed, seasoned and dressed the better-quality timbers and imported softwoods, and some cut logs, usually of the more valuable species. They commonly included a joinery shop that might make anything from windows to pulpits. They often supplied rough-sawn construction timber bought from the forest mills, and might stock builders' hardware and other materials for city builders.[27] The statistics for New South Wales sawmills illustrate some of the differences between the two types of mills (Table 1.1).

Small sawmills regime

As the industry expanded, the readily accessible, good-quality logs were cut out, or burnt in the settlers' clearing fires. The forest sawmills were forced further into the forests, and had to build tramways along which they could bring in their logs and take the sawn timber out. The tramways were extended progressively as the cutters moved inexorably on and as the mills were moved further into the forests from time to time. The first timber tramways were built in the 1840s and the basic technology lasted for almost a century, although light iron rails came to replace wooden ones, and small steam locomotives replaced horses.

Forest sawmill at Beech Forest, Otway Ranges, Victoria, 1903–04.
(John Flynn collection, National Library of Australia)

Backhouse and Anderson's small forest sawmill at Cabbage Tree Creek, Braidwood District, NSW, 1916.
(National Library of Australia)

Eventually, a whole network of tramways and abandoned mill sites scored the forests; many can still be traced.[28]

The tallest trees, prized by sawmillers, often indicated soils prized by farmers who sought to select for settlement. The sawmillers, who were not allowed to select land just for its timber, usually needed to cut the logs on a given block progressively over a period of years, whereas settlers wanted to get rid of the trees as quickly as possible, any way they could. The inherent conflict was made worse by the sawmillers' own tramways. As in the Amazon forests today, though on a much smaller scale, the 'opening up' of the forests, in this case by tramways, let the settlers in. There were plenty of conflicts between individual millers too, as each sought to grab the best patch of forest, or bar a competitor from doing so. These conflicts were fiercest where demand was greatest and resources scarcest, in Victoria. In the Wombat Forest, for example, the sawmilling firm of Crowley and Fitzpatrick claimed that their rivals were even destroying timber to prevent them using it.[29] Although the tramways used simple technology, they could cost as much as a sawmill itself; but unlike the mill, most of the investment could not be moved. Hence, as the industry expanded and investment in tramways increased, the sawmillers needed security of access to the forest resources at least long enough to recoup their investments.

The regime that eventually developed was based on two principles that have shaped forest policy to the present day. The first was that the state would reserve certain areas of the forest against selection for agriculture by creating either permanent 'state forests', or temporary 'timber reserves' to be held until the timber was cut. The second was that the state would allocate exclusive rights to cut the timber on a defined area of the public forests for a specific period, by licence to an individual sawmiller.[30] The provisions of the licences varied between colonies and were changed frequently in their first few years.

Victoria was first to provide sawmillers with exclusive rights in the forests: in 1860, it authorised licences to be issued for up to seven years, though it did little to create reserves, in spite of major recommendations from 1865.[31] New South Wales provided sawmill licences for one year from 1861, and legislated to create permanent forest reservations from 1871. Reservation proceeded quite rapidly, so that 3.4 million hectares had been reserved by 1881 and 4.9 million two years later. Queensland's timber licensing system too was expanded in 1868 to provide special timber licences with exclusive rights. These served the merchants sending shiploads of logs to the southern colonies as well as the nascent sawmilling industry. In 1875, the Tasmanian sawmillers started to press the government that selection should be kept two years behind logging to allow them time to remove

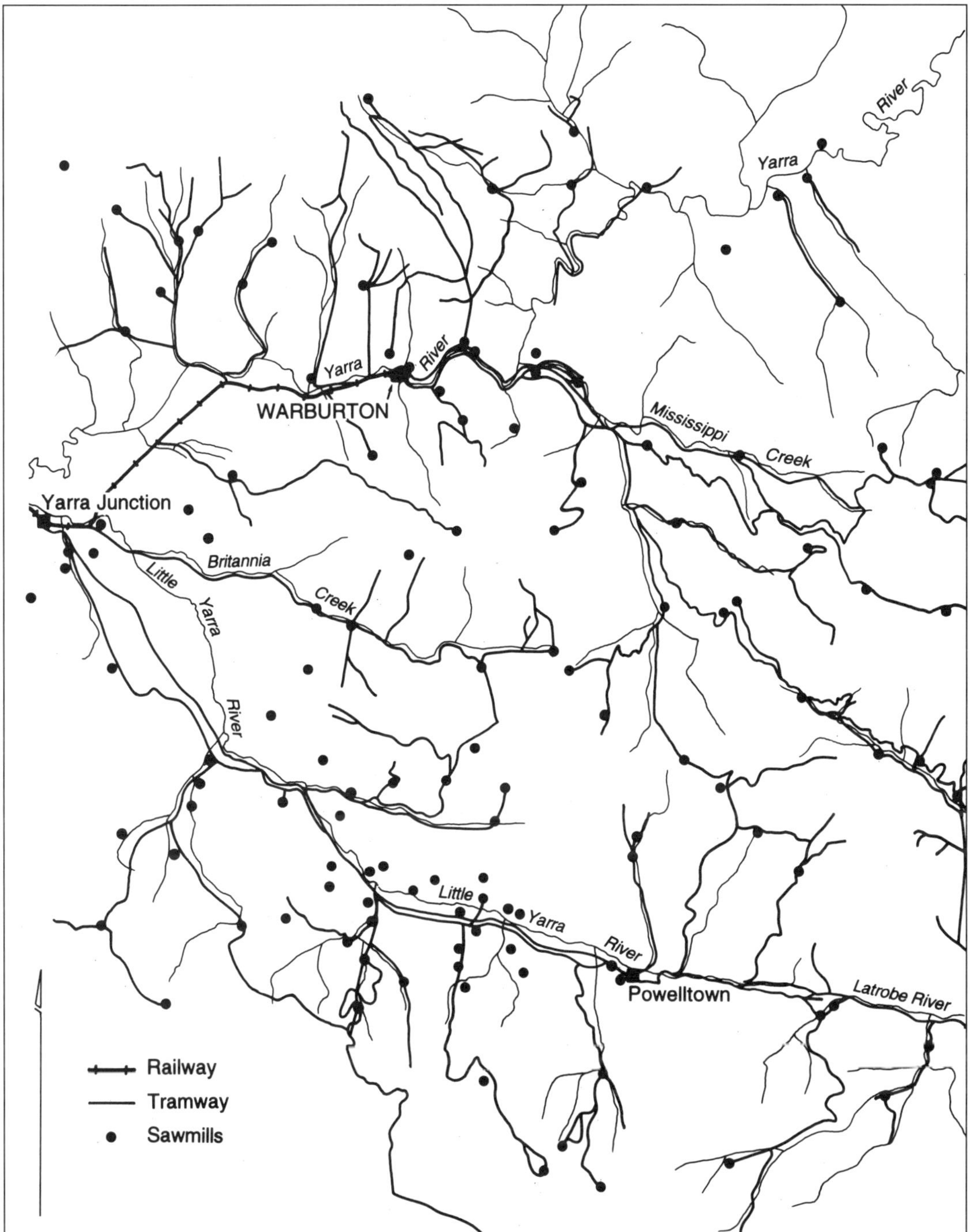

Figure 1.3 Pattern of forest sawmills and tramways around Warburton, Victoria
Source: Redrawn from part of *Powelltown District Sawmills and Tramways* map, Light Railways Research Society of Australia, 1982 (research maps 1 and 2).

the timber, but it was not until 1881 that legislation was passed to allow reserves 'for the preservation and growth of timber' and by 1884, only 10 000 hectares had been declared.

Political influence largely determined who got the licences, explicitly so in Victoria where the Minister issued them. Although they were commonly issued for periods as short as one year, their renewal or transfer became an entrenched administrative matter which provided considerable actual security.[32] And although the legislation set upper limits on the area of a licence, big firms often had more than one; for example, the big Queensland firm of Wilson Hart in Maryborough had nine or ten licences, each of a square mile (260 hectares), by 1875.

Forest bureaucracies had to be created to administer the regime.[33] These were very small sections within the powerful Lands Departments, or sometimes in Agriculture or even Mines Departments. New South Wales appointed two rangers in 1875, and by 1883 had a Forest Conservancy Branch with a chief clerk, an inspector and thirty-nine rangers and assistants to manage reserves of some 2 million hectares.[34] Victoria first tried local Forest Boards, and then a central one of three members from 1874; presumably the real administrative work was carried out by the Lands Department's district officers. In Tasmania, police officers acted as Crown lands bailiffs to administer the licensing system for manual cutting in the forest, while the Lands and Survey Department provided an often tardy central administration.

The new regime was far from fully installed or enforced. The progress of declaring forest reserves was painfully slow, and many sawmills saw the stands they expected to log selected from under their noses, or were even blackmailed when cunning selectors blocked their access. The small forest bureaucracies had very limited ability to oversee forest operations, or mark the boundaries between licence areas, so that conflicts between sawmillers were not eliminated.

Forests in the age of rail and empire

The transformation of the world system between the mid-1870s and World War I marked a new stage of capitalism; it was epitomised by the spread of railways and empires, and it added a new regime to the forests of remote Australia. Politically, the world map was redrawn as European powers scrabbled for resources, markets and territory as never before, and as small principalities gradually amalgamated into powerful nation-states. The Australian colonies too, after long argument, formed the Commonwealth of Australia in 1901. Its new 'States' retained responsibility for land and forests, while the Commonwealth was made responsible for international affairs and trade in a division of powers which now proves uneasy. Technologically, the map had to be constantly revised to show new ports, naval stations, canals and the rapidly spreading railways. Lines were built across North America by 1869 and Russia by 1903; 40 000 kilometres of railways spread across India; in Africa, lines were pushed down from Egypt to the Sudan, and from Cape Town and Durban to the interior, especially to the great new mines of the Rand. In Australia, more than 11 000 kilometres of railways were built between 1870 and 1890, and there were over 16 000 kilometres by the end of the century.

Science and technology advanced on every side. Steel became cheaper and better. For example, high-tensile, flexible steel was made into bandsaw blades capable of sawing timber faster, more accurately, and with less waste than ever before.[35] Steamships became common, the old wooden sailing ships being relegated to poorer cargoes, like timber; and steam engines were adapted to agriculture and logging. New financial institutions marshalled capital for heavy industries in Europe and exported it for railways, mines, sawmills, and pastoral properties. In the British Empire, and most industrialised nations, public administration was reorganised into efficient and impersonal systems to take on an ever-increasing range of activities. While the imperial nations extinguished rights by conquest abroad, they extended parliamentary democracy at home. In Australia, representation was widened, particularly by extending the franchise to women between 1894 and 1909. Parliaments came to have working-class representatives and by 1910 the Labor Party even won federal government. Unionism gradually extended to small and isolated activities such as the timber industry, where it improved wages and conditions, and controlled the iniquitous truck system.

Wood demand and trade

Hard, durable timber, often in large sections, was essential for the world's new infrastructure; every kilometre of railway needed 3360 wooden sleepers.[36] Wood was even used to pave the streets of London and Paris, at least in the 'better' quarters. If local forests were limited, as in Britain, sparse, as in much of Africa, or remote, as in much of India, then suitable timbers had to be shipped across the world, an enterprise that became more attractive as freight rates fell. Urbanisation, rising affluence and ornate taste in big cities everywhere demanded fine timbers from the isolated forests of many countries.[37]

Australian forests contain many species with durable timbers. The red gum forests along the River Murray and forests in northern New

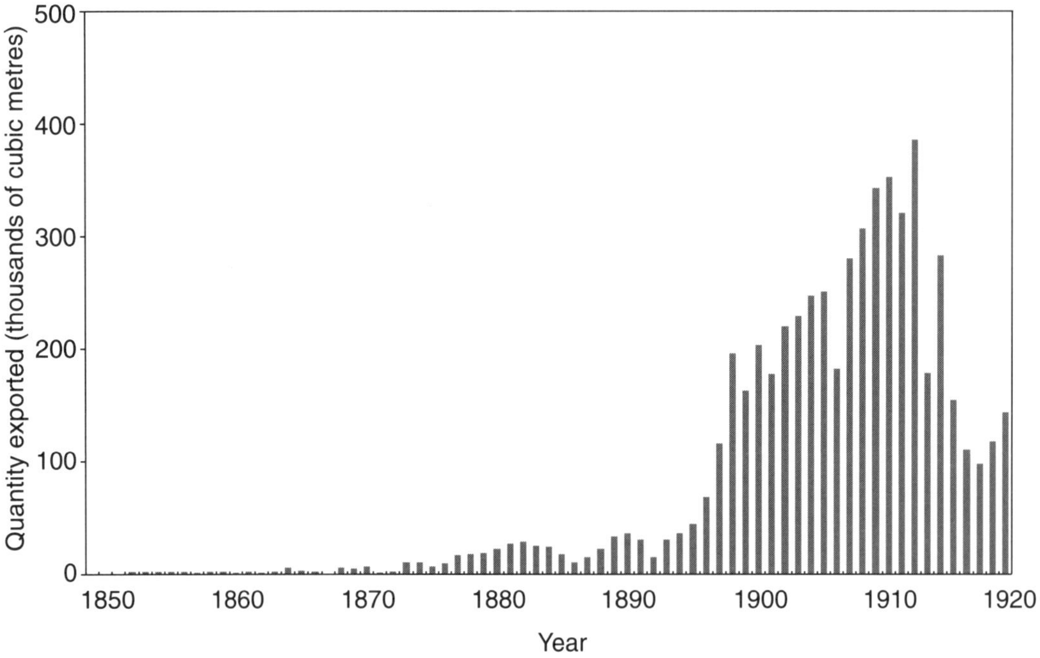

Figure 1.4 Timber exports from Western Australia, 1850–1920
Source: WA Statistical Registers.

South Wales had suitable timbers but they were used primarily within Australia. The forests of southern Tasmania supported a specialist export trade in long, straight blue gum piles which, the Tasmanians proudly proclaimed, built the docks at Dover and the Royal Navy's base at Simonstown. But it was only the jarrah and karri forests of Western Australia which had enough durable timber to support large-scale exports, and it was there that a new regime of extensive concessions to support large sawmills was developed.

Western Australia had developed later and much more slowly than the eastern colonies, so that in 1870 the non-Aboriginal population was still only some 25 000. The domestic demand for timber had been easily satisfied by manual production and a few small local sawmills. Small quantities of jarrah had been exported successfully and it was realised that the forests could support timber as an export staple if demand increased and the forests could be accessed.[38] It was railway building in South Australia and Victoria in the 1870s which stimulated the demand, and it was their entrepreneurs and railway contractors who built the first large sawmills.[39] Demand was also increased somewhat by railway building in Western Australia in the 1880s, and more so in the 1890s by the rush of people and capital to the great goldfields of Coolgardie and Kalgoorlie. However, it was international demand which caused a dramatic boom in the industry from the 1890s and made Western Australia the source of 70–82 per cent of Australia's timber exports.

Although small early sawmills operated successfully in the northern jarrah forest close to Perth, any substantial increase in production had to come from the best jarrah and all the karri forests which grew further south. Several factors differed markedly from those in the eastern colonies. The forests were remote from their markets and 30 to 50 kilometres away from sheltered points for loading ships. Cargoes had to be much larger for shipping in ocean-going vessels than for the vessels that plied the eastern coasts. There were few settlements or roads to support the timber industry. Steam sawmills with large, heavy-duty equipment capable of sawing the large trees had to be built in the forests, complete with settlements for the workers, and railways had to be built to take the timber to port. Such investments were beyond the scope of most of the small sawmill operators in the eastern colonies and could be undertaken only by groups of entrepreneurs or by companies.

Large sawmill regime

In contrast to the very limited provisions granted to sawmillers in the eastern colonies where agriculture and mining were dominant, Governor Weld in Western Australia developed a new resource regime to attract investment. Long-term leases, or 'concessions', over very large areas of forest were offered on the provision that substantial investments were made in their exploitation. Three concession agreements made with British approval in 1871 were followed by a series of others in Western Australia and two in Tasmania. From the 1920s a number of concessions were granted under which the pulp and export woodchip industries now operate.

The first three agreements in 1871 for the Canning, Jarrahdale and Lockeville concessions provided exclusive long-term rights over 100 000, 250 000 and 'about 200 000' acres (40 500, 101 200 and 80 900 hectares) at little or no rent, although licence fees had to be paid for the area worked, and for the cutters employed in the bush.[40] Other conditions varied and changed. For example, the Jarrahdale lease was for twenty-eight years at an annual rental of £50, after which it could be extended for fourteen years. Slight changes were made when it was finally granted in 1874, and when it was transferred in 1889, but its basic provision of exclusive rights for forty-two years stood. Other sawmillers were provided with areas of up to 16 600 hectares by special licence for fourteen years, but the great expansion of the industry in the 1890s was supported by issuing leases over up to 30 300 hectares for up to twenty-five years, provided 'a substantial and fully equipped plant' was constructed within two years.

Colonial capital

Although Governor Weld probably hoped to attract British capitalists, it was largely timber men from the eastern colonies who took up the first concessions, and they were probably under-capitalised for the difficulties of pioneering new areas and markets. Three notable examples are described below. It was only when their pioneering had proved successful, twenty years later, that British capital came to take advantage of the soaring demand.

A syndicate of Victorian businessmen, including Peter Lalor, formed the Rockingham Jarrah Timber Company which took up the Jarrahdale concession.[41] They imported '40 to 50 tons of machinery' for their sawmill, and built 37 kilometres of wooden-railed tramway to take their timber to Rockingham, where they built a deepwater jetty. The company was restructured in 1874 and again during the 1880s, and changed hands in 1889 and again in 1892. It replaced its wooden rails with iron, bought more steam locomotives, built further sawmills and

The Mornington sawmill, Jarrahdale, WA, about 1885.
(J. S. Battye Library)

Timber on the Jarrahdale–Rockingham line, *Clarion*, Spring 1897.
(J. S. Battye Library)

Loading jarrah railway sleepers and timber for London at Rockingham jetty, WA.
(J. S. Battye Library)

jetties, and expanded production to the extent that it was supplying 40–50 per cent of the colony's export trade. By 1896 it had four sawmills, a 'very fine planning and grooving' mill, several maintenance workshops, and 64 kilometres of tramway into the forest as well as its railway to Rockingham. Most of the 320 workers lived in the Jarrahdale settlement, which by that time was complete with store, post office, telegraph, hospital, doctor, school and church, and eventually a hotel.

The industry extended into the karri forests further south. These were magnificent indeed; the great trees towered to almost 90 metres, with long, straight boles clear of branches. The most notable development was that of M. C. Davies, an experienced timber man and railway builder from South Australia. He first had a licence for fourteen years, but this was renegotiated in 1882 to provide exclusive rights to the timber on 18 600 hectares of forest for forty-two years at a rent of £150 a year.[42] He built a jetty at Hamelin Bay big enough to load three ships with steam cranes, 24 kilometres of standard gauge railway, and two large sawmills equipped with vertical breaking-down saws big enough to handle the huge logs. By the end of the century there were 65 kilometres of railways, steam locomotives, and workshops. Two hundred men were regularly employed, and the settlement at Karridale had a store, hall, school, public library, church, doctor, and even a race track.

The two Millar brothers came to Western Australia in 1884 to build the Great Southern Railway. They constructed several sawmills and acquired 8000 hectares of karri forest which had been granted to the railway.[43] By 1896 they had two sawmills, and were in the process of erecting two more. Logs were brought in on 8 kilometres of tramway and their 'extensive establishment' was connected to the railway to Albany where the timber was loaded, largely for the London market.

British capital

Colonial firms needed British capital to expand fast enough for the market opportunities of the late 1890s. The Jarrahdale company was refinanced with British capital, and took a new name; M. C. Davies floated his company in London to obtain capital for further expansion; while Millars opened two new sawmills, and then floated the firm in London as Millars Karri and Jarrah Forests Company. All this, and the leases being offered by the government, attracted a rush of investment. By the end of the century, the nine leading timber firms had a total capital value of over two million pounds, much of it British, held leases over 390 000 hectares of forest, and owned three-quarters of the sawmills.[44] The twenty-two largest forest sawmills were big concerns which together employed nearly 3000 people, only three of whom were

Table 1.2 Sawmills in the States, 1902

State	*Number of mills*	*Average number of employees*	*Average engine power (kW)*
New South Wales	331	12	14.8
Queensland	150	12	17.8
South Australia	24	11	10.2
Tasmania	67	14	13.6
Victoria	208	16	13.6
Western Australia	27	110	89.5

Source: Statistical Registers.

women; they used 1700 horses and bullocks, and operated 460 kilometres of railways and tramways. Even when averaged with the older small and town sawmills, the Western Australian sawmills were clearly larger than those in the eastern States (Table 1.2).

Prices fell during the Boer War (1899–1902) which led eight of the British-based companies – including Jarrahdale and M. C. Davies – to amalgamate into Millars Karri and Jarrah Co. Ltd. in 1902. The 'combine' held 336 000 hectares of the best jarrah and karri forests; this virtual monopoly enabled it to increase local prices, press the government to reduce rail freights, and try to cut the wages of its workers.[45] One local firm, which much later was to dominate the Western Australian industry, Bunning Bros, survived outside the combine, largely because it operated as a local timber merchant dealing with imported softwoods, dressed timber, and building supplies, and relied less on its sawmill.[46] Millars' dominance was challenged in 1912 when the Kauri Timber Company bought a firm with two forest leases and re-established it as a substantial sawmilling enterprise.[47] Nevertheless, the industry in Western Australia remained highly concentrated.

British capital was also invested in southern Tasmania. Although manual cutters and small sawmills had exploited the forests there for half a century, their tracks and wooden tramways had touched only the coastal fringe. The rugged hinterland could be opened by railways and steam power under the new regime. The Huon Timber Company and the Tasmanian Timber Corporation were formed, and promptly received exclusive concessions of 8900 and 10 100 hectares from the Tasmanian government for forty-two years.[48] The Huon Timber Company built what was claimed to be the largest, most technologically advanced sawmill in the southern hemisphere. However, the railway and other costs were so substantial that by 1908 neither company had

Logs on the Huon Timber Company's over-capitalised, standard gauge railway system. Geeveston, Tasmania, about 1920.
(W. Beattie; State Archives of Tasmania)

paid a dividend. An energetic local company, Henry Jones and Company, was appointed to manage their affairs; it failed to make them profitable and they were sold to Millars Karri and Jarrah Company for about one-third of their capital cost.

Labour

Mechanised production demanded the organisation of labour on time wages within the sawmills. In the numerous small mills, proprietors and family members worked alongside hired labour. In the new large sawmills, managers and foremen controlled the men on behalf of distant shareholders; the mill whistle, not the owner's shout, marked the working day. Thrown together in large numbers in the new sawmills, and in the town mills and timber yards of the capital cities, workers gradually joined the rising labour movement and, helped by the Australian Workers Union, formed their own unions. This proved vital in Western Australia, where attacks on wages and hours, particularly by Millars, were resisted in a long and bitter strike in 1907.[49]

Although labour in the mills was paid time wages and became increasingly unionised, fallers in the bush were almost always paid piece-rates. The railway systems of the large sawmills were operated on wage labour, but in most of Australia, it appears that most of the bullock teams hauling in logs, and many of the horse teams and carts taking timber to market, were owned by individuals operating as contractors to the sawmills.

Paradoxically, manual production, in which the workers were paid piece-wages, expanded concurrently with large-scale sawmilling, especially in cutting railway sleepers from the jarrah forests of Western Australia.[50] Sleeper cutters fashioned the forests differently to the sawmills. To make any sort of a living they had to choose the most suitable trees to fell and logs to split. By comparison to the sawmills, whose steam power could reduce even the most cross-grained log, they had to be more selective. The objective conditions in which they lived and worked need to be seen behind the enlarged photographs that now romanticise their lives. Typically, a couple of workers, or perhaps half a dozen, camped in the bush in rough huts of poles and bark slabs, moving from place to place as needs be. In Western Australia, camps of up to 500 people, including women and children, were housed in timber-framed huts, clad with galvanised iron, that could be dismantled for moving to new sites. Hessian bags covered the windows, and what furniture there was came from slabs of wood and old kerosene tins. This regime enabled merchants and timber companies to avoid the costs of building sawmills and providing the very modest level of social infrastructure for them. Millars extended their attack on workers by cutting the piece-rates they paid for hewn sleepers by 16 per cent in 1906 and deducting haulage costs. The concentration of workers in large camps, and the newly elected Labor government, enabled the workers to form a co-operative, the South West Timber Hewers Co-operative Society Ltd. It secured orders in the face of the Millars combine, obtained a forest licence, and built its own sawmill.

Even though they led isolated lives, forest workers had a strong sense of identity. It was expressed in axemen's contests held from the 1870s, which evolved into sporting associations.[51] The Australian Axemen's Association, formed in 1891, was presided over by an ex-Premier and architect of federation, Sir Edward Braddon. Shaking off the old 'convict stain' that haunted Tasmanian timber workers, the axemen took a strongly loyalist stance. 'Nowhere has his Most Gracious Majesty [Edward VII] more faithful and loyal subjects than those comprising the society of axemen scattered throughout Australia', declared the association's manager.[52] And when the Duke and Duchess of York visited Tasmania, having opened the first Commonwealth Parliament in 1901, the axemen offered prizes and employed 'a couple of good bushmen' to collect 'beautiful berries' from the Nietta district for presentation with an address of welcome.[53]

Fashioned by use and Empire

During the period to World War I, the forests were used in specific regimes which facilitated production and capital accumulation. Most notably, the state behaved differentially: large British capitalists were encouraged to invest with the allocation of large concessions; small colonial firms were given only limited rights against the opposition of agriculturalists; and individual producers were treated as subjects for control rather than help.

The main interest in conserving the resources of the forests was a short-term one: to save stands of good timber from clearing and settlement for long enough to enable the trees to be harvested rather than burnt. The patterns which this period marked on the forests reflected the technology and economics of the day; they were mostly a mosaic of small areas cut year by year for small mills. They spread from the coasts and rivers, later from railways and roads, along tracks or tramways back into the forests, usually bypassing steep or inaccessible places. The Western Australian pattern, cut for larger mills on flatter country, was somewhat coarser. Within each patch that was cut, the men took only the best-quality trees of sizes that could be handled. The number and types of trees that were left varied widely: in the rainforests of North Queensland, only the most select logs of the most valuable species were taken, as few as a dozen trees per hectare; in the mountain ash forests of Victoria and Tasmania, almost every tree was merchantable. But it was not only cutting which made the pattern so intricate; the vagaries of nature governed how many young trees of which species replaced those felled, while random and repeated bushfires, or cyclones in the north, stamped their patterns over all. The urgent fashioning of Australia's naturally varied forests created an intricate pattern. It was little suited to provide the mass of uniform raw materials required for the industrialisation which was to follow.

CHAPTER TWO

Forests for Industrialisation

Australian forests could hardly escape being fashioned in new ways when the whole world order was changed by the two world wars and the great depression between them. Then, in the 1950s and 1960s, the old empires crumbled, colonies became nations, and, Bob Menzies notwithstanding, Australia lost its British genuflexion. Expansion was spectacular: industrial production soared, prodigious scientific advances were flaunted from the moon, a billion more people were added to the world. About 2.5 million refugees and immigrants entered Australia; this influx combined with the high birthrate to generate an intense demand for houses, and for the timber to build them.

Consumption rose rapidly in what came to be pejoratively labelled as the 'consumer societies'. Australia was certainly one of them, Japan was another, and paper a product which epitomised the label. Not only was more paper consumed for traditional uses such as printing, writing, books and newspapers, but supermarket shopping used more packaging, and new 'personal care products' were devised for every human orifice. Consumption was met by establishing large pulp and paper mills, whose voracious appetites for wood changed Australia's forests more radically and on a larger scale than any before. This chapter describes how the surging Australian demands for timber and paper fashioned the forests with two new regimes; Chapter 4 describes how the Japanese demand did so later.

Australia's forests were used as part of a development strategy adopted after World War I to counter the lack of manufacturing, especially in heavy industry, which had been revealed. The new strategy aimed to modernise and diversify the economy by increasing agricultural exports, processing them further, and expanding manufacturing to replace imports. Governments knew too that something had to be offered to the soldiers returning with expectations of a better life. They began the new strategy with a will. Economic institutions, such as the Tariff Board and the Loans Council, were created, and in 1920 the Institute of Science and Industry was set up to develop new technologies – it became the Council for Scientific and Industrial Research (CSIR) in 1926 and the Commonwealth Scientific and Industrial Research Organisation (CSIRO) in 1949. The public infrastructure of irrigation, electrification, roads, and telephones was expanded, though not always efficiently, with British capital, creating public debts which Australia rued during the great depression of the 1930s.[1]

Capital and initiative came from both British and Australian sources, particularly from Broken Hill Proprietary Limited (BHP), and the loose 'Collins House' grouping of Anglo-Australian mining companies and old pastoral capital. It also came from some smaller Australian firms, but these needed to amalgamate before they could afford to enter heavy industries such as pulp and paper making. Although plans were prepared, many investments were made only when the depression eased at the end of the 1930s. A strand of company welfare ran through these plans. For example, Gerald Mussen persuaded Collins House that 'mere money-grubbing' should give way to a 'Gospel of Happiness', consisting of company provision of health care, education, decent housing and more than subsistence wages, all of which were applied when three major pulp and paper mills were built.[2]

There were two major possibilities for industrialisation in the forests sector: if eucalypts could be seasoned properly, then perhaps they could replace some of the imported softwood timber; and if they could be pulped successfully, then they could supply enough raw material for making paper. These possibilities were economically important because, by 1920–21, Australia was importing £4 million worth of softwood timber a year, and almost as much printing, newsprint, and wrapping paper; other wood products brought the total import bill to £12 million. By contrast, timber exports amounted to only £1.5 million.

Establishing the pulp and paper industry

Paper had been made mechanically in Australia from rags, recycled paper and imported pulp since 1868, but expansion required wood pulp. However, the technologies which had been developed for northern hemisphere conifers did not suit eucalypts. The Institute of Science and Industry set up a small Forest Products Laboratory in 1919 which worked closely with industry to develop appropriate methods. By the late 1920s, both printing paper and newsprint had been made from eucalypts in this laboratory, but commercial success was far from clear.[3] Large-scale tests were then made in overseas mills, and Collins House even built a small experimental pulp mill in southern Tasmania in 1928, only to close it after the Wall Street crash in 1929.[4] Three major groups of capitalists weighed the uncertain possibilities throughout the 1920s and 1930s, and jockeyed for position, especially for the newsprint market. The depression delayed investment, but the structure which emerged at the end of the 1930s left each group with its own mill and a sector of the market.

One group was formed in the newspaper industry, which had concentrated during the 1920s into four large companies and a number of smaller ones.[5] Two of the largest, the Herald and Weekly Times and the Fairfax company, formed Australian Newsprint Mills (ANM). In 1937 ANM reached a secretive agreement with Canadian mills to share the Australian newsprint market on condition ANM bought some of their softwood pulp to mix with its eucalypt pulp. This entitled ANM to supply its owners' newspapers, and six smaller ones, with enough newsprint to warrant building a pulp and paper mill.[6] The Tasmanian government agreed to provide a quarter of the paid-up capital and a large forest concession, while the Commonwealth agreed to provide a bounty for the first four years of the mill's operation. With these in place, the company built the Boyer pulp and paper mill near New Norfolk in Tasmania which finally started to make newsprint in 1941.

Another group emerged when the two largest existing paper makers amalgamated and, in 1926, took over the two remaining manufacturers to form Australian Paper Manufacturers Ltd (APM). It produced most of Australia's wrapping paper and paperboard from wastepaper and imported pulp. Overseas tests, and trials at one of the company's paper mills, showed that a strong kraft pulp could be made from eucalypts that would, if blended with a proportion of softwood pulp, make good wrapping paper and paperboard.[7] In 1938 APM built a pilot-scale pulp mill at Maryvale in Victoria for further research and development; the Victorian government provided a forest concession, and the company proceeded to build a full-scale pulp and paper mill there in 1939 and 1940.[8]

As ANM and APM were clearly well entrenched in their markets, the Collins House group took up the smaller market for printing and writing papers by floating Associated Pulp and Paper Mills Ltd (APPM), which built a pulp and paper mill at Burnie in north-west Tasmania in 1936. In addition to the major mills, a group of South Australian investors formed Cellulose Australia Ltd which built a very small mill at Millicent in the south-east of South Australia in 1941. It used wood from nearby pine plantations to make paperboard by a chemi-mechanical process, and cellulose for munitions during World War II.

Expanding the pulp and paper industry

World War II shut down the flow of imported paper, enabling the new mills to sell every tonne they could make, regardless of its quality. By the war's end, they were firmly established and had greatly improved quality by research and experience. Currency controls continued to limit imports from the United States and Canada into the 1950s. Meanwhile, domestic consumption boomed so fast that Australia was consuming a million tonnes of paper a year more by the end of the 1960s than at the end of the war. ANM, APM and APPM expanded as fast as they could, some smaller companies tried to enter their markets, and two American multi-national companies entered the new market for disposable tissue products. Production increased even faster than demand, and imports were reduced to about one-third of consumption, a proportion which remained roughly the same from the 1960s into the 1990s.

Newsprint

ANM expanded newsprint production in two steps, each of which had to be negotiated carefully with national and international interests. The first added a second paper machine to the Boyer mill in 1953, which doubled its capacity. In doing this, ANM replaced its contract to buy pulp from Canada with one to buy it from New Zealand Forest Products Ltd (NZFP). To supply the pulp, NZFP too had to expand and needed ANM to guarantee its loan from Australia's biggest insurance company, the AMP Society.[9]

Meanwhile the New Zealand government entered into a joint venture with capitalists there to set up the Tasman Pulp and Paper Company in 1956 to make newsprint for export, potentially on the Australian market. They engaged the world's leading newsprint company, the multi-national Bowater Paper Corporation, to manage the mill and sell its products. In 1964, Tasman doubled its production and Bowater apparently orchestrated another division of the Australian market between Tasman and ANM. The resultant oligopoly was sealed by exchanging directors and \$2 million worth of shares.[10] ANM persuaded the other two major newspaper companies, News Ltd and Consolidated Press, to take major shareholdings, thus giving each major newspaper a stake in domestic manufacturing and a link to New Zealand in an 'orderly' market. But restraint on ANM's production did not suit the Tasmanian government which was pushing its policy of hydro-industrialisation for all it was worth. Indeed in 1964, the Minister even threatened that the government might 'reconsider its attitude to

ANM's pulp and paper mill at Boyer, near New Norfolk, Tasmania, 1972. The major expansion completed in 1968 is on the left.
(Australian Newsprint Mills Pty Ltd)

the [forest] concession', described later, unless ANM expanded.[11] ANM promptly announced its second step and opened a third paper machine in 1969, which doubled its production again.

Packaging

The expansion of the large domestic market for the many sorts of wrapping paper and 'cardboard' which APM made was due partly to the growth of manufacturing and packaging for supermarkets (which then used stout brown paper, not plastic, for shopping bags). Another factor was the substitution of paper for other forms of packaging; notably multi-wall paper sacks for jute and cotton, and cardboard boxes for wooden in Australia's fruit sheds. All this was APM's domain.

APM used recycled waste paper in most of its products and this provided up to half of its fibre. For the other half, the company expanded its Maryvale pulp mill in several stages and built a small pulp mill at Port Huon in southern Tasmania in 1962.[12] For this it adapted a new continuous neutral sulphite semi-chemical (NSSC) process for making pulp, and developed a novel way of pressing the pulp into pellets that could be stored in bulk and loaded into ships by conveyor for use on the company's Sydney paper machines. APM created a network of paper mills by expanding those in Sydney and Victoria, and building new ones in Queensland and Western Australia. They were fed from its pulp mills, recycled paper and some imported pulp.

The successful start-up of Cellulose's small mill and its operation in the post-war period encouraged a group of timber firms to start a small pulp and board mill near Launceston in northern Tasmania.[13] Delays in obtaining the machinery meant they could not start until 1953, by which time the immediate post-war chance to enter a starved market had gone. The mill closed after eighteen months and sold its board machine to APM. Cellulose was not well funded for expansion and its chemi-mechanical pulping process was outdated. In the early 1960s APM took over some of its shares and marketed its products, before taking it over completely in 1969 and later closing it.

Smorgon's entry

Although APM partly captured its market through long-term customer agreements, it was not safe from a lively, well-funded entrant. Moreover, it carried the considerable overheads of running its national network, maintaining an active research laboratory and, as described later, developing its own plantations. A number of its machines were old; it produced a large number of paper grades, many in small quantities; and its efficiency was not high.[14]

Smorgon Consolidated Industries Pty Ltd, a wealthy and energetic, family-directed firm in the meat industry, set up a second-hand paper machine in Melbourne in 1959 and commenced making wrapping paper from waste paper and imported pulp. It had negligible overheads, no pulp mill, and made only a few grades of paper which it sold at 15 per cent below whatever APM charged. It added a NSSC pulp mill to its Melbourne site in 1967, and then a second, but modern and highly automated, paper machine. By 1971 it was producing over 90 000 tonnes of paperboard and corrugating paper. The pulp mill drew its wood from state forests in the Otway Ranges west of Melbourne, from sawmill chips and from private property to the north and north-east. APM's directors were so incensed by competition that they tried to starve the new mill out of existence by sending their foresters to buy all the sawmill chips and privately owned forests they could ahead of Smorgon's people. But it was pollution that closed the pulp mill. Its chemical waste was discharged to Melbourne's sewers, rather than recycled in an expensive recovery plant, so that Victoria's Environmental Protection Agency eventually forced it to close.[15]

Printing and writing paper

APPM not only increased its production of fine paper by almost five times during the 1950s and 1960s, but added to its range of papers, and started saw, hardboard, and particle-board mills. Its own research enabled pulp production to be increased by 40 per cent in 1949; it was increased further in 1956 and 1964 when two continuous digesters were added. The company added two machines for making printing and writing paper at Burnie, three small machines for making food wrapping papers and parchment, and built a mill at Ballarat in Victoria to coat some of the papers made at Burnie. Backed by the Collins House group, it had little difficulty in raising the capital it needed, although British companies were initial partners in two of the ventures.

As the company started to use its own Tasmanian forests in 1946 it came to 'believe it to be essential to utilise every bit of timber that can be profitably turned into account'.[16] To do so it built a hardboard mill in 1951, a large sawmill in 1954 and a particle-board mill at Wesley Vale, near Devonport, 60 kilometres to the east of Burnie in 1963.[17] In 1961 APPM announced that it would build a big new pulp and paper mill there and the Tasmanian government scurried to provide another forest concession, described later. However, an economic downturn and a fall in profits made the company decide to install only a paper machine at first and supply it with pulp from the Burnie mill. Construction eventually started but it was not until 1970, when otherwise it would have lost its new concession, that paper was finally produced.

Wiggins Teape

APPM's near monopoly in making printing and writing paper in Australia was challenged in 1956 when the British paper company Wiggins Teape, with a minority interest from another British paper maker, built the Shoalhaven paper mill near Nowra in New South Wales. It was a small mill with an initial capacity to make only 8000 tonnes of paper a year, mainly from purchased pulp, as it had no pulp mill to make its own. An expansion to 30 000 tonnes a year was in train when Wiggins Teape was absorbed into the giant multi-national British American Tobacco Company; it soon sold the Shoalhaven mill to APPM, in exchange for shares amounting to 25 per cent of APPM's capital.[18] It was a high price to pay, but retained the printing and writing segment within APPM's control.

Tissues

Paper that could be thrown away replaced cotton for babies' bottoms and everybody's noses. Rival brands of such 'personal products' were promoted on television and packaged appealingly for the supermarket shelf. This was a new business for Australia that required new manufacturing and packaging technologies and new marketing strategies; technologies, brand names and consumer marketing that were well developed in the United States.

APM and Cellulose formed a joint company, Apcel, and built a pulp and paper mill at Millicent which used the sulphite process to pulp wood from the region's pine plantations.[19] The paper was converted by the American Kimberley–Clark Corporation and sold under its 'Kleenex' brand name. Cellulose withdrew, a thermo-mechanical pulping process was added, and the mill was expanded as a joint venture between APM and Kimberley–Clark. The second venture, Bowater–Scott Australia Pty Ltd, was established by the American Scott Paper Company and the British-based Bowater Paper Corporation. It operated a converting plant in Melbourne, and marketed its products under the 'Scottie' brand. In 1971 it built an integrated pulp, paper and sawmill at Myrtleford in north-east Victoria, which used pine from government plantations.

Raw material

The great expansion of paper production rested primarily on making pulp from Australian trees, although pulp continued to be imported and much more waste paper came to be used. The escalating demands for wood were met by increasing the rate of felling and using more types and species of trees. This was possible only because the companies and CSIRO continued their research and development work vigorously. When the three big pulp mills first started, they depended largely on the mountain ash forests for high-quality, pale wood with relatively long fibres. As demand increased, the mills had to use more and more of the darker wood with shorter fibres from the mixed-species foothill forests. Existing processes were improved, especially in recycling the chemicals, and new processes were devised so that even old trees from the foothill forests could be used. Softwood pulp with its long fibres was needed to give tearing strength to papers made from eucalypt pulp and waste paper, and to make tissue products. Rather than import it, the mills started to use pine from plantations. Most of the wood came from public forests, part came from land being cleared for agriculture and other private forests, and an increasing part came from public and private pine plantations (Figure 2.1).

Exclusive concession regime

The most obvious way to provide the new pulp mills with so much additional wood was to grant them exclusive concessions over large

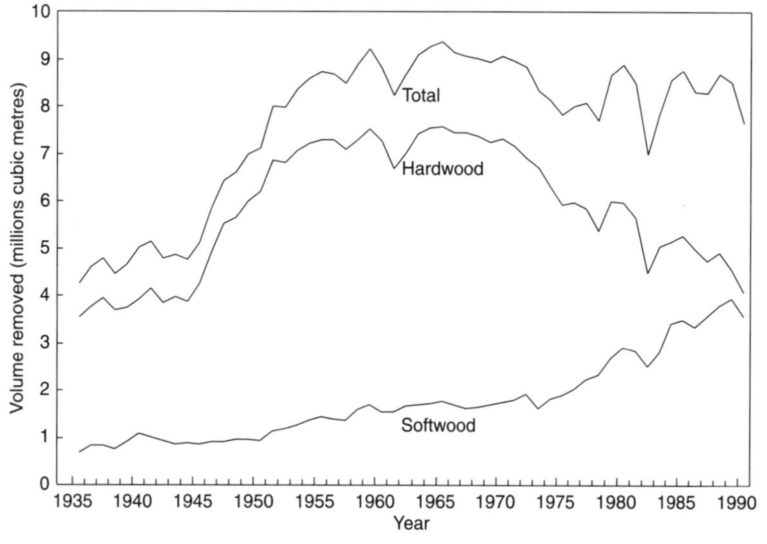

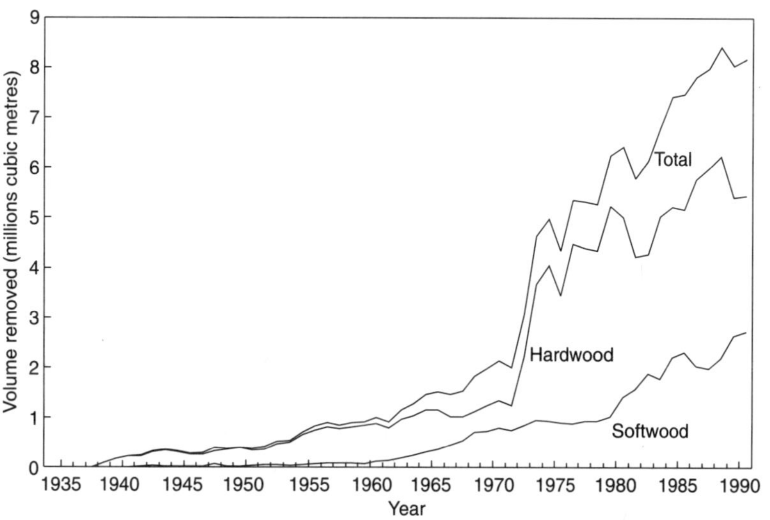

Figure 2.1 Wood from Australian forests and plantations, 1936–93: top, sawlogs; bottom, pulpwood
Source: Wilson (1969); Forestry and Timber Bureau, *Annual Reports.*

areas of public forests. Tasmania and Victoria hastened to do so by passing special 'encouragement' Acts. These established a regime similar to the large sawmills regime discussed in the previous chapter, but provided greater security and coverage. Although the States retained ownership of the forest land in each case, the companies obtained rights and responsibilities over the wood resources.

The concession granted to APPM at Burnie in north-west Tasmania originated from negotiations in 1924, well before the mill started in 1936. As the forest had not been even surveyed then, the concession was simply declared to include the Crown lands from 16 kilometres north of Zeehan and within 24 kilometres of the Emu Bay railway. Exclusive rights were provided to all the pulpwood and the company could cut the sawlogs too, but not exclusively so. Surprisingly, no period was specified: the rights apparently still exist in perpetuity.[20] No rent was charged, but the company had to pay a very small royalty which could be increased as its profit rose.[21] A quarter of the concession could be dedicated as permanent forest to be managed by the state; who was to manage the remaining three-quarters was not specified but, almost by default, management devolved to the company. Certainly the Conservator of Forests, L. G. Irby, believed that it would do so on the grounds that 'a pulp-wood company must practise sound forestry principles if it is to have a prolonged tenure of existence'.[22] To the public resource, the company added 100 000 hectares of forest land it bought from the old Van Diemen's Land Company.

The second concession was granted to ANM in 1932 over the magnificent forest of the Florentine Valley. The resource of approximately 120 000 hectares was particularly attractive, not only because of the giant size of the trees, but also because it contained a high proportion of mountain ash, the most favoured species for pulping. However, the inaccessible location and huge trees meant that heavy investments had to be made in engineering a route into the valley and in the equipment to handle the logs. The company was granted exclusive rights, for eighty-eight years, to use both the pulplogs and the sawlogs for whatever it willed, provided that it did not overwhelm the local sawmilling industry by selling more than 20 per cent of the wood on the sawn timber market. The small sawmillers operating on the edge of the concession had their rights terminated, but the company was obliged to supply them with 20 000 cubic metres a year of logs. ANM, like APPM, was required to pay a small royalty on the wood cut, according to its profitability. Although the Forestry Department had gained more experience by 1932, it was still far too small to manage such a large development. The total management of the Florentine forest was devolved to the company who had to regenerate the logged areas, protect it from fire, and prepare a long-term management plan approved by the Conservator.

Victoria, unlike Tasmania, had a vigorous sawmilling industry operating throughout most of its public forests, and a well-established Forests Commission. It already had management plans specifying where cutting could take place and how much could be cut each year, and even if poorly funded was clearly in charge of protection and regeneration. The concession granted to APM over forests in the central highlands of Victoria provided exclusive rights, for fifty years, only to the pulpwood.[23] The rights were intricately defined in schedules of the 'minimum annual supply' quantities that the Forests Commission had to make available from various qualities of forest.[24] It was envisaged that APM would cut its pulpwood from the trees or heads that the sawmillers left behind after their logging operations had finished. As sawmillers were operating in many different locations, the pulpwood would have to be procured from many small, scattered logging operations. However, APM had not even started its main pulp mill when the terrible 1939 bush fires ravaged Victoria. They burnt 1.4 million hectares of forest, including nearly all of the prime ash forests in the concession. The loggers turned to salvaging the burnt areas, but although the mountain ash, a fire-sensitive species, was killed, the trees were so large that it took twenty years for decay to render them unusable for pulp.

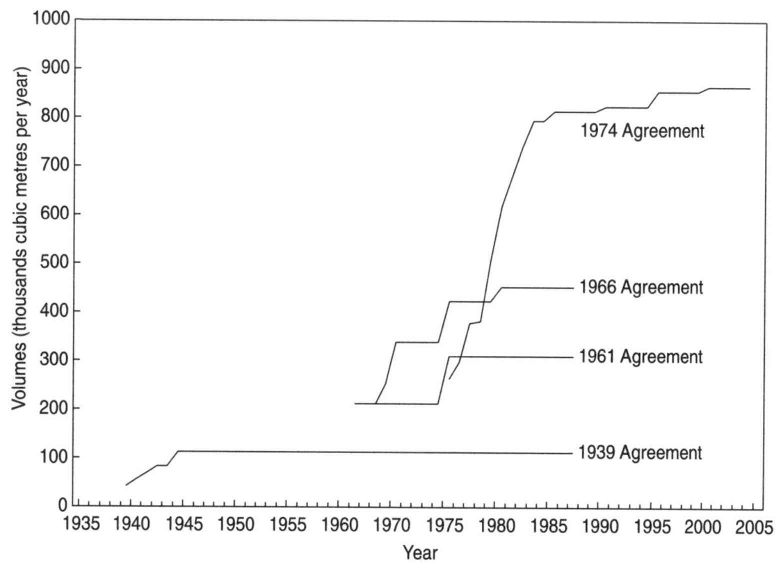

Figure 2.2 APM's minimum annual supply quantities in Victoria
Source: Victoria, *Wood Pulp Agreement Act 1936, Forests (Wood Pulp Agreement) Act 1961, Forests (Wood Pulp Agreements) Act 1974.*

APM expanded rapidly in the 1950s and 1960s and was able to have its forest concession progressively enlarged and the available quantities increased. Its wood came from many different types of operation dispersed across the region. Most came from sawmilling areas; some was from areas, such as Boola Boola State Forest described later, which were logged primarily for their pulpwood; and roughly one-third came from land being cleared for agriculture and pine plantations.

Although the exclusive concession regime started the pulp industry, it could not support its expansion; there were no more big, vacant forests, like those in Tasmania to allocate, and the sawmills needed more resources too. Before describing how a new regime was devised to integrate the two industries, the increasing pressure of sawmilling on the native forests needs to be examined.

Sawmilling and panel manufacturing

The technological changes envisaged in Australia's strategy of industrialisation and import replacement could be introduced into sawmilling only gradually. The small, family sawmills, which made up the bulk of the industry, had little capital for improvements, and the industry was depressed for much of the time between the wars. It was in full production briefly to meet the demand for housing after World War I, but an economic downturn in 1922 and a flood of imports depressed production. It did not recover before the great depression drove markets and production down and down. The small forest sawmills battled as best they could, frequently working short time, paying short wages, laying off men, or too often closing altogether. The export-oriented mills in Western Australia were severely affected by the loss of their international markets during both the wars, as well as by the depressions. There was little money to pay for new equipment; the old had to do.

Nevertheless, the possibility of developing better methods to season Australian timbers so that they could replace imported softwoods was pursued successfully in CSIR's laboratory. By the early 1930s it had improved the design of kilns to dry sawn timber artificially and, most importantly, had invented a method of 'reconditioning' it to overcome the internal collapse of eucalypt wood on drying. CSIR sent its officers to mills in every State where they advised on the installation and management of kilns, and reported problems that needed further research. By the end of the 1930s, the techniques for kiln drying were well developed and the products established in the trade. Although the cost of kilns was beyond the reach of most of the forest mills, a few large

Timber workers march to Darlinghurst in Sydney during the 1929–30 strike. The twelve men named served prison sentences of up to two years.
(Australian Timber Workers Union, NSW Branch; Noel Butlin Archives Centre, Australian National University)

sawmillers and timber merchants installed them, and some centralised kilns were built to serve a number of smaller mills.

No 'Gospel of Happiness' was preached by Australia's sawmillers. Isolated workers in the forests and small forest mills had to do as best they could with the boss they had and whatever house, hut or shack was there; sometimes they did reasonably well, sometimes miserably. Petty chiselling workers out of their full wages was widespread in the smaller mills, but in the big mills and city timber yards the larger numbers of men were less easily intimidated. The struggles for better wages and conditions erupted into two major, long-lasting and disastrous strikes. Each tried to defend a 44-hour working week granted under a federal award in 1920, each was fought on a falling market, and the Timber Workers Union lost in each case. The first, in 1921–22, was started by Huon Timber Company in southern Tasmania, which engaged a contractor to hire scab labour for some railway realignment work. The workers went on strike, the company locked them out, the contractor's men were ambushed, and their boarding house was blown up. The bitter strike lasted for a year and yielded only a pyrrhic victory for the workers' suffering; the mill closed the next year, and the employers persuaded the Arbitration Commission to put the hours of bush work back to forty-eight. In 1929 the employers secured the Lukin award, which set hours for the remaining timber workers at forty-eight, and reduced other conditions throughout Australia. The union in Tasmania was too weak and dispirited to strike, but in New South Wales and Victoria, it conducted a long and bitter strike only to be beaten in the end.[25]

Recovery from the depression occurred only gradually from the mid-1930s until the start of World War II. Then the drop in imports combined with military building to create such a demand that the mills had to press their mostly worn and run-down equipment to full production. In the immediate post-war period, Australia could meet little of the intense demand for timber by imports, and domestic production had to be increased. The first and most obvious response was to add slightly over 1000 sawmills to the 1700 that existed at the war's end. Most of these new mills were tiny affairs: 655 of them had four or fewer workers, and consisted of nothing much more than a war-surplus engine driving a single circular saw. They were cheap and could be moved readily to wherever a few logs could be found or scrounged, particularly from patches on private property which were easiest to get to. Builders were so desperate to get timber that they had to accept its poor quality.

Many of the seventy-one people who had lost their lives in the terrible 1939 fires had been burnt to death in mill settlements deep in the forests. For safety in future, these mills were relocated to towns or townships at the forest edge, but there were other factors too. Economically, relocation was possible because it cost little to cart the logs a few extra miles by road. Technologically, electricity offered better control and greater convenience than steam or diesel engines, but meant that mills had to be near the electricity grid. Socially, workers wanted better access to community facilities and education for their children; few were prepared to live for long in the rough, isolated sawmilling settlements that had been so common. With full employ-

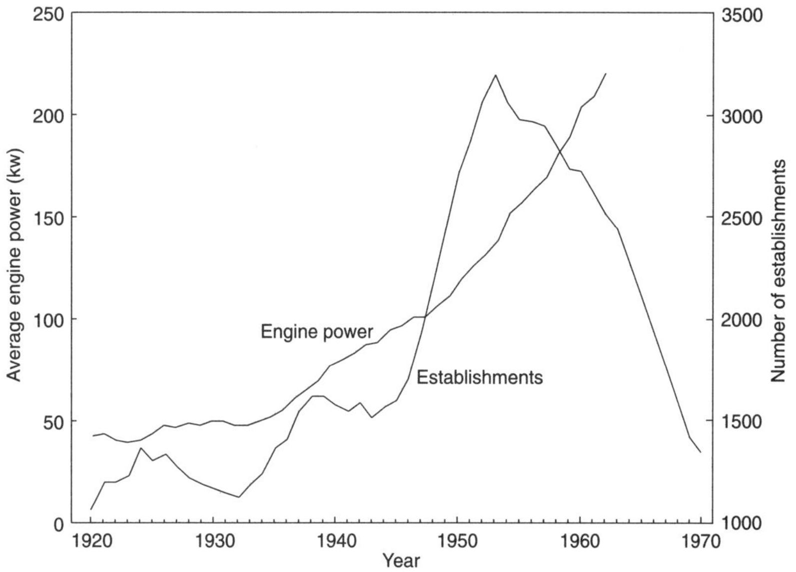

Figure 2.3 Sawmills, number and average engine power, 1920–70
Source: CBCS, *Secondary Industries Bulletin*; ABS, *Manufacturing Establishments: Details of Operations by Industry Class* (Cat. no. 8203.0).

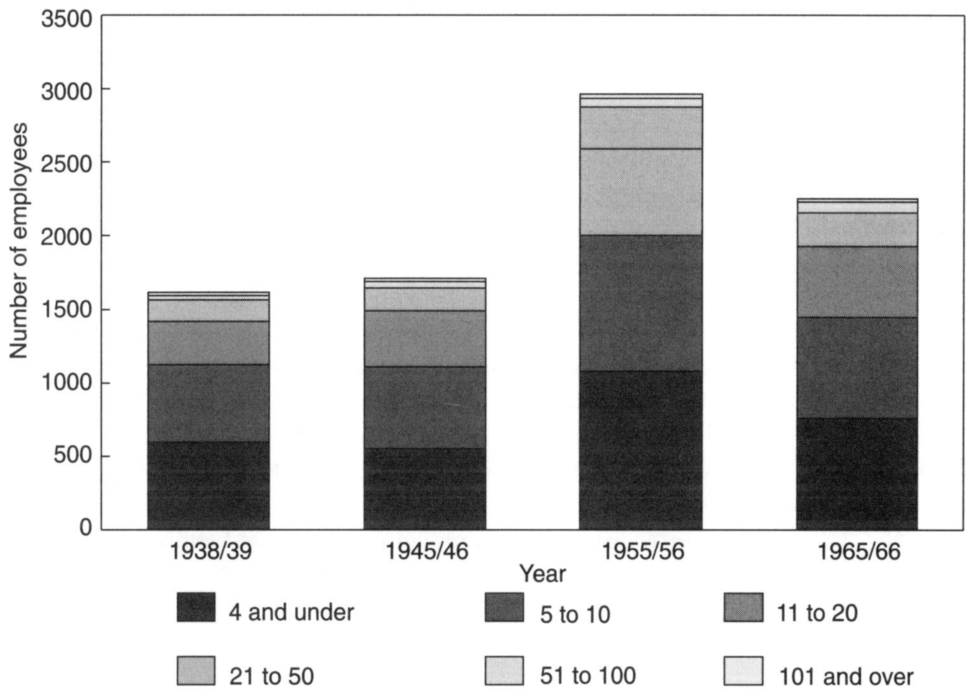

Figure 2.4 Number of sawmills by employment size class, selected years
Source: CBCS, *Secondary Industries Bulletin*; ABS, *Manufacturing Establishments: Selected Items of Data Classified by Industry and Employment Size* (Cat. no. 8204.0).

ment and a more accessible, more concentrated labour force, the union was able to have the conditions in the award improved, although it proved to be a long hard slog for the organisers to get them fully implemented in the small mills.

Large mills equipped with new machinery and kilns were built from the mid-1950s, and as more of these came into production, the number of tiny mills dropped away. Part of the reduction was due to APM's success in persuading fruit and vegetable packers to install automated packing lines which used cardboard rather than wooden boxes. Nevertheless, a core of tiny mills remained, many working part-time to supply local demands for farm timber. Sawmills were also built to cut the increasing quantity of logs becoming available from pine plantations, whose concentrated and uniform logs were better suited to industrialisation than those from the hardwood forests. Although some small pine mills were built in all States, mostly to produce case or other low-grade timber that could be sold air-dried, the major new pine mills were larger, more technically sophisticated, and more highly automated than the largest hardwood sawmills. Equipped with drying kilns, dressing and stress-grading machines, and packaging sheds, they could produce carefully seasoned, accurately graded, and well-presented timber for the market. Half of the resource was in South Australia, where for historical reasons the government, not companies, was the major sawmiller.[26] By the end of the 1960s, pine provided one-sixth of Australia's sawn timber production, and by the 1990s one-half.

Sawmills were able to sell their waste – about one-third of their log input – as woodchips wherever they were near enough to the pulp mills, or the new particle-board mills described later.[27] The economic radius for sales grew to 300–400 kilometres for the best hardwood chips, and to 600 kilometres for the best pine chips. Income from chip sales was important during cyclical downturns in the timber market, and as the quality of hardwood logs worsened. Small mills, cutting less than about 8000 cubic metres of logs a year, could not afford to install a chipper, and carting their waste to a central chipper was rarely satisfactory, so that woodchip sales became an additional factor in the concentration of the industry into large mills.

The transition to larger, more automated mills was repeated in making panels. The rising demand for cheap doors, furniture, kitchen and bathroom 'vanity' cabinets, television cabinets and so forth led to an increased use of panels where solid timber had been used before. The three main types of panels – plywood, hardboard and particleboard – were produced by successive introductions of new technologies into Australia.[28] Mills to cut veneers and make them into plywood had long been established as relatively small businesses, much like sawmills, in the major forest regions of the east coast, and in Brisbane, Sydney and Melbourne. They required top-quality logs and faced many of their panels with Australia's most decorative woods from the rainforests. But the highly automated mills used thermo-mechanical processes which could make first hardboard and then particle-board from poor-quality hardwood or small sized pine. Ten particle-board mills were built by some of the large timber merchants and by major companies such as APPM, APM and CSR. Their clean and sophisticated factories presented a modern industrial image which contrasted sharply with that of the hardwood sawmills and ply mills.

The mill regimes

The small and large sawmill regimes retained their essential feature of allocating defined areas of public forests exclusively to particular mills. However, the regulations were altered as demand increased, resources became scarcer, the forest services gained more control, and industry concentrated into larger, more expensive mills. To raise revenue and reduce wasteful cutting, the States had introduced royalty payments to be made on the quantity cut each year, but cheating by the sawmillers on their payments was endemic. As one vigilant inspector was to comment in 1937, 'there never had been a month when all the [sawmiller's] log sheets [on which payments were calculated] have been correct'.[29] Regulations proliferated in attempts to control the problem. As road transport of logs became common, the problem of illegal logging became serious too. Some States required millers to hold a licence to operate a sawmill, in addition to any licence they held in the forests. Mill licences were required in New South Wales from 1916, in Queensland from 1936, and in Victoria from 1939 for mills in, or within one mile (1.6 kilometres) of, state forests, primarily as a fire control and safety measure.

As the forest services became better established, they adopted the principle of regulating the amount being cut from the forests by attaching maximum annual quantities to the forest licences. In Queensland the maximum productive capacity of each mill was specified on its licence and Ministerial approval had to be obtained to enlarge or move a mill. The forest services often set the maximum quantities on the production capacity of the existing sawmills, rather than on the growth capacity of the forests. As they became better able to calculate the growth capacity, they attempted to reduce the amount being cut by reducing the licence quantities 'pro-rata' across all the mills in a particular zone. Generally the reductions were too little and too late, which was to have drastic consequences for the industry in the 1970s and 1980s.

An intricate relationship between the sawmillers and the forest services evolved. The administration of the increasingly comprehensive licence regulations became more and more important in determining the resource rights, and hence the sawmillers' profits. There were three notable features. First, although the regulations allowed licences to be issued for periods as short as one year in some States, they were renewed or transferred so automatically as to provide long-term security for investment. Indeed, expectation of renewal was so entrenched as to create an economically tradeable value, although it was found to have no legal value when tested in the New South Wales courts in 1980.[30]

Second, the system functioned to favour some groups of millers over others. Two classes of hardwood sawmills were created: 'Crown' mills which had their licences on state forests renewed regularly, and 'private' mills which drew most of their logs from private land, supplemented with occasional parcels of logs from state forest. As the forest services did their best to keep 'their' Crown mills supplied, new entrepreneurs were shut out of the industry unless they could take over an existing Crown mill for its quota of logs, or could develop ways of using poor logs that the main mills would not take. Forest licences with their log quotas were transferred to new owners when sawmills were sold, so that take-overs became the primary means of concentrating production into larger mills. However, the provisions of the sawmill licences, or the royalty charging system, were sometimes used to prevent the relocation of production. This was so in Victoria during the 1950s and 1960s, when forest licences were let only to mills in their region of the State. Queensland had a particularly fussy system of zoning, which like Victoria's ostensibly encouraged the decentralisation so loved by the Country Party, but which also functioned to protect the large sawmills on the coast from competition from inland mills.

The distinction between the resource rights granted to large and small mills re-emerged, despite the ending of long-term concessions for large sawmills granted under special Acts when the Huon Timber Company closed in 1923 and the early grants to Western Australian mills eventually expired. Various mechanisms were used in the different States, including special licences, special agreements, and agreements ratified by special legislation. They provided exclusive rights to sawlogs or pulpwood for between ten and fifty years, and commonly specified rights and responsibilities for the provision of infrastructure. These mechanisms were used not only for a few of the large sawmills, but also for the new hardboard and particle-board mills.

Integrated regimes in the hardwood forests

The rapid expansion of sawmilling and pulp-making in the 1950s and 1960s necessitated a new regime. Somehow sawlogs and pulpwood had to be found from the same forests and the exclusive rights of different mills had to be accommodated. What emerged was an integrated regime managed by a much higher level of state planning.

The first approach separated the timing of the two logging operations. Sawmillers who were used to picking through their licence areas every few years were told to cut all the sawlogs on each patch, after which other contractors would fell all the remaining trees for pulpwood. This created conflicts and inefficiencies. The sawmillers who built the secondary roads resented the pulpwood contractors who paid nothing to use them later. However, as the sawmillers' roads were only lightly built for dry-weather use, the pulpwood contractors who used them throughout the year often damaged them, or the pulp companies had to pay to upgrade them for heavy traffic in wet seasons. As the sawmillers' fallers were paid piece-wages, they cut only trees which were certain to have sawlogs in them, and left those about which they were uncertain or which were not worth the effort of cutting for only a small sawlog. Thus the pulpwood cutters, who clear-felled virtually every merchantable tree, encountered some potential sawlogs which were cut up for pulpwood. Moreover, the naturally regenerated seedlings which often came up after the sawmillers' logging operations were in many cases badly damaged by the subsequent pulpwood operation.

The second approach was to have only one logging contractor who cut both products in one 'integrated' operation, sending sawlogs to the sawmill and pulpwood to the pulp mill. It was first tried by APM in 1948 in the Boola Boola State Forest, 30 kilometres north of its Maryvale pulp mill.[31] The area had been repeatedly picked over by sawmillers, who had left few sawlogs. The company obtained exclusive rights to log everything that was left; it brought in heavy equipment, built substantial all-weather roads, and progressively clear-felled the forest. The logs were graded by Forests Commission supervisors, who directed an average of 85 cubic metres per hectare to the Maryvale pulp mill and 15 cubic metres to local sawmills. Notably, the sawmillers would not have been able to obtain these sawlogs by the first approach.

The integrated regime was developed much further by the Tasmanian Forestry Commission in the 1960s in order to supply APM's Port Huon pulp mill. It rested on what was then the most thorough inventory of the quantity of the logs in the resource and was expressed in two detailed plans.[32] It recognised that the forests were honeycombed with sawmillers' permit areas which should not be upset. Hence, every logging operation had to produce both sawlogs and pulpwood concurrently. Some of the small sawmills carried out their own logging directly, while the large ones employed contractors to do so. These produced and sold pulpwood to APM. APM too engaged contractors and, much as at Boola, sold sawlogs to the sawmillers. The Forestry Commission could order loggers to conduct integrated operations and could settle disputes as to costs; in practice, integration was achieved by persuasion and costs were settled by negotiation. Although the plans regulated both APM and the sawmillers, the latter were consulted only 'as necessary for the purpose' in the formal planning process conducted by a committee of Forestry Commission and APM officers.

Integrated logging gradually spread around Australia to wherever more than one product was obtained, and to private as well as public forests. The forest services supervised the operations to prevent sawlogs being cut into pulpwood, and occasionally the other way. In some areas, such as Boola, the logging was conducted primarily to produce pulpwood, with sawlogs being produced almost as a by-product, but in most it was planned primarily for sawlogs.

Fashioned by new technology, high demand and heavy industry

It was not until after World War II that the forest landscape was radically transformed by new logging technology, the escalation of felling, and the new regimes. The age of rail and steam, of small bush mills and isolated settlements dotted through the forests, which had inscribed the landscape for a century and more, had ended; the age of road and truck, of pulp mills, electricity and towns had begun. The transformation was radical because it made virtually every hectare and every tree accessible; exploitation was simultaneously extended and intensified.

Bulldozers drove roads far into the mountains, opening up resources unattainable before. The mountain forests of the alpine and East Gippsland regions of Victoria were made accessible for the sawmills which had lost their resources in the 1939 bushfires. And rough fire tracks were pushed across the most remote areas, so that at last crews could get to bushfires in less than a day. Bulldozers could also sneak into spots that tramways could never have reached, and with their winches tweak logs from the most awkward places. Chainsaws made cutting so much easier that trees left before became feasible to fell. This was important because most sawmillers had to get their logs only from

Roading the forests with bulldozers opened up steep country previously inaccessible to tramways. Nightcap National Forest, Lismore, NSW, 1937. (Forestry Commission, NSW, *Annual Report*, 1937)

trees left or grown in areas that had already been logged. To get enough, they were forced to take progressively poorer qualities of logs, until only those fit for pulpwood, or too few to be worth harvesting on their own, were left.

The forests which supplied the pulp mills were changed radically by clear-felling. Rather than select trees with sawlogs here and there and leave the rest, as the sawmill regimes had, most integrated logging operations felled almost all. Thus, the stands which had been an intricate jumble of trees of all ages were replaced by even-aged crops of young seedlings growing on the areas cut over. Although not necessarily with that intent, these forests were being transformed to provide a much more uniform raw material for an industrial future. But, there could be no long-term prospects unless the forests were conserved and protected, and their resources renewed; neither could they delight the Australians who visited them unless they were preserved and cared for. How these concerns also fashioned the forests is the subject of the next chapter.

CHAPTER THREE

FORESTS FOR CONSERVATION AND DELIGHT

The fashioning of Australia's forests was described in the previous two chapters as driven first by immediate colonial needs for land and wood, then by an export trade in durable timbers, and from the 1920s by a strategy of industrialisation and import replacement. But from the middle of the nineteenth century, the waste and depletion of resources were increasingly countered by measures to conserve the forest for future use. At the same time, their values for delight and recreation, protection of catchments, and the amelioration of climate were asserted, which led to some being reserved from exploitation. This chapter traces several strands in conservation up to the 1960s, their incorporation into the resource regimes, the rise of professional forest services to manage them, and the creation of national parks and similar reserves. As with the economic and political dimensions, a mix of international and national interests was involved.

Case for forest conservancy

Three strands can be traced in the start of forest conservation: the gross waste and damage to the forests spread rapidly, the rising scientific community argued what came to be called 'the case for forest conservancy', and the value of the forests and their products became much more widely known.

The grossest assaults on the forests occurred on the Victorian goldfields where perhaps 100 000 people cut almost every tree around them. What was not used was wasted or burnt. In 1865, three senior officers reported to the Victorian Parliament that:

> A great number of persons employ themselves cutting timber for the mines. They fell the best, destroy more than they use, consequently there is unnecessary waste. Persons resident on the gold fields destroy also an immense number of fine trees, by taking sheets of bark from them for roofing their huts. ... We have ascertained that one very great agent of destruction is fire.

> It is certain that if steps be not taken to enforce the more economical use of native timber, and to conserve the forests, there will soon be difficulty in getting timber for the uses of the miner, in any forests adjacent to the gold workings.[1]

They recommended that a large forest reserve should be created, extravagant waste prevented, and plantations commenced, but a decade later it was clear that little had been done.[2] In every colony there were similar, if smaller scenes. Queensland's Colonial Botanist, Walter Hill, reported that:

> The immense timber forests of the colony, containing every description of wood, many of very great value, would, if properly looked after and judiciously conserved, be a great source of national wealth and even revenue, ... but instead ... the most reckless waste and needless destruction of our most valuable trees has been permitted to take place year after year with impunity ... their extinction has not been caused by legitimate and profitable consumption ... but entirely through the wholesale destruction of trees by timber-getters, settlers and others, through ignorance of their value.[3]

After hearing Hill's and other evidence, a Select Committee recommended prohibiting log exports, creating forest reserves, supervising timber-getters, setting up a small advisory 'Forest Conservancy Board', and starting a small forest tree nursery attached to an asylum.[4]

Colonial Botanists, such as Walter Hill, were particularly important in Australia because agriculture and horticulture relied entirely on the introduction of exotic plants, and immigrants sought the comfort of familiar flowers. The first botanic garden was established in Sydney in 1816, and by the 1870s they could be found from Ballarat to Rockhampton. Of several notable government botanists, Ferdinand Mueller, appointed by Victoria in 1852, was outstanding.[5] Doctors, clergymen, magistrates and such people collected and collaborated with Mueller and other government botanists, and formed learned societies to encourage science.[6] Their endeavours were integrated into the international scientific community through extensive individual correspondence, the interchange of journals and herbarium specimens.[7]

As the government botanists continued to introduce exotic plants, they found that softwoods from the northern hemisphere could be grown successfully. In 1870, the director of the Botanic Gardens in South Australia, R. Schomburgk, called for the establishment of plantations to counter the 'wasteful destruction' of that colony's very limited forests. A Forest Board was established in 1875 to oversee

Forest devastated by gold miners at Forest Creek, Golden Point, Victoria, 1858.
(Richard Daintree; La Trobe Collection, State Library of Victoria)

reserves, and a Woods and Forests Department in 1883. Trials of various species conducted for quarter of a century showed that radiata and maritime pines could grow particularly rapidly and might be able to overcome Australia's reliance on imported softwood timber.[8]

Technology and applied science were celebrated in European exhibitions, museums and spectacular architecture. The newly self-governing Australian colonies included timber among their displays in British exhibitions and anxiously persuaded the Admiralty's timber inspectors to examine them. As science advanced, the inspectors adopted systematic, quantitative tests, so that by 1875, Thomas Laslett could report on the strengths of woods from around the world, including six Australian species.[9] Science and technology were also celebrated in Australia. A small museum in Hobart was the first to open in 1841, followed by museums in Sydney, Adelaide, and Melbourne by 1857, and Brisbane in 1877.[10] Exhibitions were staged in Melbourne in 1867, and Sydney 1870, but it was not until Sydney staged an International Exhibition in 1879 that much timber was shown. Among the produce that New South Wales displayed was a 'most magnificent, extensive and instructive collection of 320 specimens of the various [indigenous] timbers'.[11] Queensland sent a 'magnificent and extensive collection of 260 varieties of timber', while Victoria and numerous private exhibitors sent varieties of timbers 'beautifully shown'. From the

Forest Creek in 1994, naturally regenerated to forest and used for rough grazing.
(John Dargavel)

1890s, scientific tests of the strengths of the various timbers were conducted in Australia, and these became commercially important as the export trade developed.[12]

The exhibitions and museums were popular with the public, and the natural history displays attracted great interest. Natural history associations, with a largely suburban membership and an interest primarily in wildlife, were formed from the 1880s.[13] Widespread interest in Australia's plants and animals was expressed in art and popular journals, and facilitated for many city people by cheap excursion fares on the extending railways. The populace flocked to picturesque spots for picnics, while the sturdier went walking through the bush – the first of many walking clubs was formed in 1894 – and couples took their honeymoons in guesthouses in forested places around the cities, such as Bundanoon (NSW) or Marysville (Victoria). Governments responded by reserving such spots as parks, considering them 'a national recreation and pleasure ground ... for the amusement, recreation and convenience of the inhabitants'.[14] In 1879 an area of 73 000 hectares was reserved in the sandstone country south of Sydney as the Royal National Park, an exceptionally large size for the time. By the end of the century, popular interest in forest botany was such that the New South Wales government published an attractively illustrated flora in which 'scientific accuracy will in no way be sacrificed to homely language'.[15]

Greater understanding was also needed of the overall quantity of the forest resource. Locally, experienced timber men continued to locate the stands to last their own mills the next few years. But determination of the overall resource was complicated by the infinite variety of the forests, few of which had been surveyed accurately. The initial estimates could easily be grossly wrong. For example, the first Conservator of Forests in Western Australia, John Ednie Brown, travelled where he could through the jarrah and karri forests and, in 1896, estimated their area at 8 million hectares; a quarter of a century later they were found to be only about one-sixth of that.[16] By 1916, only South Australia and Victoria had forest maps, and these only on a large scale. Even when the areas could be identified, the quantities of timber on them still had to be determined. The president of the Royal Society of New South Wales, J. H. Maiden, urged that trees should be measured to find out 'the climates and soils in which various species flourish best' and how fast they grew.[17] He complained in 1908 that 'in not one of the States ... have we a proper forest survey – stock sheets of the contents of the national forests'.[18]

More urgent was the need for clean water in the cities. The early water supply schemes for rapidly expanding Melbourne, for example, drew water which was contaminated by human excrement, cattle trampling and erosion on the banks of the feeder streams.[19] By the 1870s damage to the catchment by the loggers and paling splitters led to a public inquiry which received evidence from Mueller and Robert Ellery, the Government Astronomer. It recommended that cutting be stopped, not only in the catchments themselves but in the 'crests and brows of the timber range', thought to be 'very important ... as tending to increase rainfall and condensation from low clouds'.[20] Typhoid deaths continued in Melbourne during the 1880s, probably due to contamination of the reticulation system, which encouraged the government to close the catchments to people and logging.

Emergence of state forestry

From the 1880s, the case for forest conservancy was argued more forcefully. Notably, it was seen as the responsibility of the state; a Queensland Minister was even to describe forestry as 'that nationalised form of farming for which Governments are held peculiarly responsible'.[21] In this, Australia followed the imperial model developed in India and Africa: the 'demarcation' of the best forest land, dispossession of the original owners, ownership and control by the state, and management by a cadre of professionally trained foresters.

The first Australian foresters were recruited from Germany.[22] In an era when Australian farmers were clearing trees as fast as they could, this imperial model appeared as the only remedy; private forestry was inconceivable.

Those arguing the case in Victoria arranged a visit by a Conservator of Forests from India, F. D'A. Vincent, in 1887. His report was so scorching of the government's ineffectiveness that it was not printed.[23] However, the government did appoint George Perrin as the first Conservator of Forests within the Lands Department.[24] He managed slowly to build up a small staff, organise training, start nurseries and plantations, and make recommendations for the reservation of state forests. However, he was kept in a subordinate position in the department, and the major issue of controlling the exploitation of the native forests was side-stepped. In 1895 the government was persuaded to invite no less an imperial personage than the Inspector-General of Forests in India, B. Ribbentrop. He noted that the importance of forests was 'generally well understood' and 'widely acknowledged', but he castigated the government because:

> the forest laws ... are inadequate ... the area of inalienable State forests has not been increased ... the protection of the forest against fire has never been attempted, and neglect and waste in their treatment are now ... rampant ... and the money spent on their protection, maintenance, and improvement is entirely inadequate.[25]

He laid the blame squarely on the politicians of the day and 'the disregard of the general public weal where this clashes with the monetary profits of individuals or classes who can exert a direct Parliamentary influence'. The Victorian government could hardly ignore such a pukka sahib, but it avoided possibly unpopular decisions by appointing a plodding Royal Commission. By 1901, it produced

Table 3.1 Employment of forest field staff, 1912

State	*Number employed*
New South Wales	90
Queensland	9
South Australia	26
Tasmania	1
Victoria	68
Western Australia	32
Total	226

Source: Hutchins 1916, p. 180.

fourteen reports endorsing much of what Vincent and Ribbentrop had said.[26] Even so, it was not until 1907 that the Forests Act to establish a separate Forest Department was finally passed. In 1910, a forestry training school was set up at Creswick to staff the new service.

Not to be outdone, New South Wales conducted its Royal Commission in 1907, but although the government legislated promptly in 1909, it kept its Forest Department within the Agriculture Department, and left the leasing of forests in the Lands Department. This changed in 1916 when it formed the Forestry Commission. Queensland too did not provide its small Forest Department with a distinct status, preferring to keep it within the Lands Department until 1906. Western Australia had its Royal Commission in 1903, but did little until 1914 when it engaged a former Conservator of Forests from southern Africa, D. E. Hutchins, to prepare a report; this led to a substantial Forests Act in 1918 and a Forests Department.[27] Tasmania was last to do so in 1920.[28] The limited extent of state forestry before World War I can be gauged from the small number of field staff employed by the various departments (Table 3.1) to manage more than 4.5 million hectares of forest reserves.

The first step of reserving land as state forests for the permanent production of timber proceeded with varying degrees of hostility from the dominant agricultural interests and Lands Departments. By World War I, Victoria had permanently reserved some 1.6 million hectares. New South Wales had reserved 2 million hectares, of which only 0.8 million were reserved permanently in 1916. Queensland had about 0.4 million hectares, though with little legislative protection. Tasmania had few reserves for state forests and these were not necessarily permanent. Western Australia, where timber was so important, was still locating the forest areas and had no permanent timber reserves. South Australia had only about 60 000 hectares reserved, but these seemed sufficient for its new plantations.

Land continued to be reserved for a few, mostly small parks intended primarily for the delight and recreation of the populace.[29] The important parks movement in the United States of America had some influence on the trend and established the label 'national' for parks reserved by colonial and later State governments. In Queensland, the grazier and MLA, Robert Collins, was impressed by the reservation Yellowstone National Park when he visited the United States in 1896. He returned to advocate reserving part of the McPherson range (now the Lamington National Park) in order to preserve its scenery and scientific value, as well as to provide recreation; an area of 9100 hectares in the Bunya Mountains was dedicated as a park in 1908. The last State to establish a national park was Tasmania which reserved the

Mount Field park of 11 000 hectares in 1915. While the dedication of state forests and national parks proceeded slowly against opposition from agriculturalists and Lands Departments, they were not in conflict with each other. Indeed, in Queensland the forest service managed both from its creation in 1906 until a separate parks service was created in 1975.[30]

The cause of forestry

Federation encouraged a national perspective. In 1911 the Directors or Conservators of Forests from various States, and the Government Botanist from Tasmania, gathered in an interstate forestry conference for the first time. They adopted 'the object of endeavouring by every means in our power to further the cause of Forestry in Australia'.[31] They emphasised the need for special forest legislation, permanent and inalienable forest reserves, and permanent management authorities with statutory powers.[32] A second conference in 1912 started to consider the relationship of present resources to future demand, but with only crude estimates of Australia's forest area, could only compare the area of forest per head of population with that in France and Germany.

After World War I the 'cause of forestry' was advanced on a much higher plane as part of Britain's attempt to shelter its economy within the Empire. If the Empire was to be self-sufficient, then forestry would have to be practised, not merely talked about. The argument was advanced by loyalist appeal, social status, endowing 'forestry' with moral virtue, and government measures. From 1920, a series of Empire Forestry Conferences were called to discuss policy, and an Imperial Forestry Institute was formed in Oxford to disseminate technical information and to give advanced training to forest officers of the Indian, colonial and dominion services. An international cadre of cheery, beery foresters evolved with a widely shared set of beliefs, values and friendships. These were strengthened and expressed by forming the Empire Forestry Association in 1922, which sought 'a constructive forest policy whereby the sylvan resources of the Empire may be scientifically conserved and prudently exploited for the mutual benefit of the British Commonwealth of Nations'.[33] The Empire connection was emphasised in Australia through Sir Ronald Munro-Ferguson, a British politician with impeccable connections and an interest in forestry, who was Governor-General from 1914 to 1920. He returned to Britain in 1920 and, as Viscount Novar, became the first chairman of the Empire Forestry Association. He maintained a personal correspondence with C. E. Lane Poole, a forester of the British colonial tradition *par excellence*, who became the first Conservator of Forests in Western Australia in 1916 and the first Commonwealth Inspector-General of Forests.[34] Intelligent, highly trained, experienced, energetic, forceful, intolerant and perhaps arrogant, Lane Poole pressed the cause of forestry with vigour.

The imprimatur of Empire and such elite connections bolstered the cause of Australian forestry. The Interstate Forestry Conference of 1920 set an aim of 10 million hectares to be reserved as state forest, which was quickly endorsed by the State Premiers, and reported at the Empire conference.[35] As the 1920s progressed, the Commonwealth set up the Forestry and Timber Bureau for research and information, and the Australian Forestry School for training professional foresters, both ruled by Lane Poole.

The existing resource regimes were based on the state acting primarily to facilitate the exploitation of the forests by providing enough protection for the sawmillers' capital; they would not do. Sufficient state forests had to be reserved; the quantity which could be cut would have to be regulated so as not to exceed the potential growth rate of the forest; and funds would have to be found to protect the forests from fire, regenerate the areas cut over, tend the young trees, and establish plantations. Notably, the returns from these activities could not be expected for thirty to one hundred years. Not surprisingly, there was considerable opposition. The States were far less cooperative with Commonwealth organisations or each other than conference

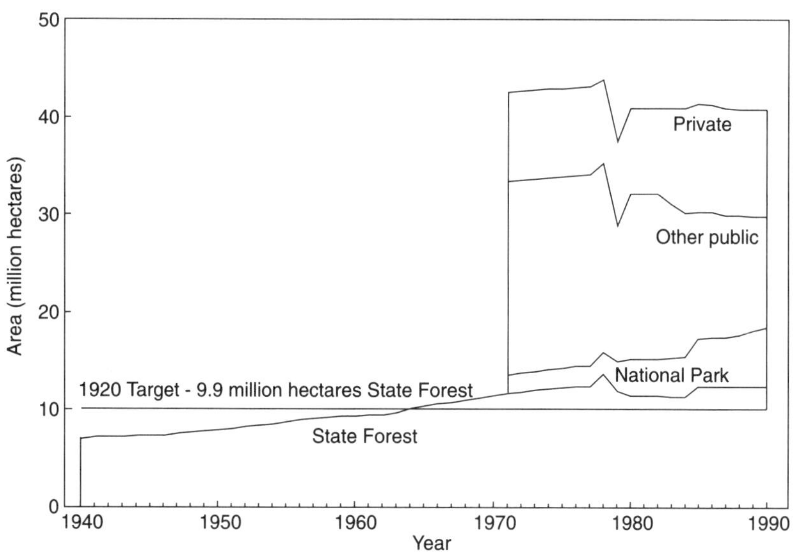

Figure 3.1 Area of forests by tenure, 1940–90
Source: Wilson (1969); *Australian Forest Resources*.

reports suggest. The sawmillers were quite opposed to having their production regulated; they feared the costs of forestry, and fought any extension the forest services which might end their cheating on royalty payments. In Tasmania, the Conservator, S. W. Steane, lamented that 'any mention of control or restriction was regarded as heresy of the most dangerous order ... it was only as such opposition was overcome that the [forest conservation] policy could be formally recognised by the Government'.[36] The imperial model of forestry assumed that an independent, dedicated professional bureaucracy would protect the forests against the 'evil influence of politics'.[37] Perhaps valid at the time, such arrogance, still redolent of its imperial origins, was to prove disastrous in the 1970s and 1980s.

Fashioned by conservation

The cause of forestry was slowed by the wars, the great depression of the 1930s, and the opposition. The reservation of state forests proceeded slowly, so that the national target agreed in 1920 was not reached until 1965 (Figure 3.1). As the foresters saw it, delay was due to obstructive Lands Departments, but they had considerable problems in selecting suitable forests without adequate maps and in still unsurveyed regions, such as north-west Tasmania. In practice, the timber of the forests on Crown land outside the dedicated state forests was also controlled by the forest services, so that industry regarded them as part of their long-term resources. Indeed, the Forests Commission of Victoria even included them in its plans for APM's long-term expansion (see Figure 2.2).

The slender forest services had to control loggers, collect revenues, protect forests, regenerate logged areas, and treat stands to improve their productive capacity. It was a tall order. During the depression, the services also provided rough camps and relief work in many labour-intensive tasks to unemployed men. In Victoria for example, relief work doubled the Forests Commission's budget in 1930–31 and enabled it to employ more than 5000 men during the year, local men getting one month's work, and city men two. It thought that 'although the service given has not reached the efficiency of that of regular forest employees, the Commission is satisfied ... that the State will have obtained much value'.[38] The services also had to survey and map the forests and prepare proper management plans. It was an enormous task involving running strip survey lines, 200 metres apart, through the forests. In the same sample year, more than 100 000 hectares were surveyed and mapped in Victoria, and working plans were prepared for almost as many, virtually doubling the area under formal plans.

Bushfire

Prime mountain ash forest burnt by the 1939 fires, Thompson River, Victoria, October 1940.
(APM Forests Pty Ltd)

Although the forest services did the best they could, bushfires could overwhelm the forests, and tragically, the people too at times. Fire is a natural part of the Australian environment, to which eucalypt forests are adapted, but its frequency and severity are significantly changed by how people start and control it.[39] Fire was essential for settlers to clear their land, but all too frequently it escaped their boundaries; fires burning sawmill waste could easily escape too, while sparks from steam engines chugging through the forests were another source; and lightning in summer storms set remote forests ablaze. Some capacity to reduce the damage from fire was an essential component of forest conservation.

The forest services built towers from which to watch over the forests, and fought what fires they could get to in time. But great fires during major droughts were beyond their ability to control; they still are. Major fires had swept across Victoria in 1851 and there were widespread fires in 1886, 1898, 1906, 1913 and 1919, but it was in 1939 that the greatest damage and loss of life occurred; seventy-one people died and sixty-nine forest sawmills were destroyed.[40] Then several small fires, which had been burning uncontrolled in inaccessible mountain areas for some weeks, met in a great fire storm which swept across the Central Highlands and many other parts of the State.

> The speed of the fires was appalling. They leaped from mountain peak to mountain peak, or far out into the lower country, lighting the forests 6 or 7 miles in advance of the main fires. Blown by a wind of great force, they roared as they travelled. ... Such was the force of the wind that, in many places, hundreds of trees of great size were blown clear of the earth, tons of soil, with embedded masses of rock, still adhering to the roots; for mile upon mile the former forest monarchs were laid in confusion, burnt, torn from the earth, piled one upon another as matches strewn by a giant hand.[41]

Judge Stretton's report as Royal Commissioner showed that such fires were man-made and that many things had to be done. The state-wide culture of carelessness with fire had to be changed, the forest sawmills had to be moved to safer areas outside the forest, settlers' fires had to better controlled, and the fire crews had to be able to get into remote areas to fight fires before they became uncontrollable. This meant that the Forests Commission had to be given enough funds and take clear responsibility for fire control. Other States had less harrowing experiences, but all recognised the need for better fire prevention and control. Doing much about it, however, had to wait until better times after World War II.

Silviculture

Not only did the forests have to be reserved and protected, they had to be 'cultivated' to ensure that new trees were grown and the resources renewed. The task was difficult in most of the eucalypt forests because many of the trees were old, scarred by fires, twisted or too defective for good sawlogs.[42] The sawmillers cut the best and most favoured species, though often returning many years later when forced to turn to poorer logs or search for trees grown larger. The poorest and least favoured trees left in the forest inhibited the regeneration and growth of a new crop, thus reducing its productivity well below its potential.

Hutchins was unfair in 1916 when he complained that 'there is no conception of the "cultivated" forest' in Australia,[43] in fact, the Department of Forestry in New South Wales had been treating areas for natural regeneration since at least 1911, and 'ringbarking useless timber' since 1915.[44] However, there were few funds – occasionally boosted by relief and prison farm workers – for removing unmerchantable trees, and it took a considerable while for the forest services to become firmly enough established to control forest operations. Silvicultural systems to deal with the situation developed with the professionalisation of forestry from the 1920s, and were applied where markets permitted. In Victoria, for example, the young

Silvicultural working gang, Jiliby Forest, NSW, 1916.
(Forestry Commission, NSW, *Annual Report*, 1916)

Silvicultural ringbarking of unmerchantable trees, like that in foreground, with natural regeneration three years later. Ourimbah Forest Reserve, NSW. (Forestry Commission, NSW, *Annual Report*, 1915)

stands coming up on the forests devastated by goldmining could be thinned for pit props under silvicultural prescriptions, and experiments in 'hardwood improvement cutting' were started by the Victorian School of Forestry in 1932.[45] In Western Australia, Lane Poole experimented with thinning karri in 1919–20, sending its wood to the Forest Products Laboratory to see if it could be made into pulp, while New South Wales produced a pamphlet; *Sylviculture*, about the same time.[46] Scientific knowledge about the eucalypts gradually accumulated, marked in 1955 by the publication of Max Jacobs' *Growth habits of the eucalypts*.[47] With more liberal financing after World War II, various 'timber stand improvement' operations were conducted to remove unwanted trees, but their value was uncertain and most of the forests were untreated. When the possibility of exporting woodchips appeared in the late 1960s, it presented a chance to dispose of otherwise unwanted trees on extensive areas and seemed to be a most desirable solution to a long-standing problem.

Plantations

The case for forest conservation also included the idea of replacing imported softwood with locally grown timber. The development of

Mature radiata pine plantation aged forty-eight years, Mt Gambier, SA. (John Dargavel)

productive plantations on almost useless coastal dunes and heathland in the Landes region of France was held up as an example for Australia.[48] It had the particular appeal to governments that coastal dunes were not sought for agriculture, but led to many of the earliest plantations in Victoria, Tasmania and Western Australia being established on dunes which were so poor that they failed.[49] South Australia was more successful and more energetic. Radiata pine was selected as the main species to be planted in the higher rainfall regions of the Adelaide Hills and the south-east region, near Mt Gambier. To produce sawlogs, it was envisaged that the trees would be grown on a rotation of forty years with thinning at intermediate ages. By the start of the twentieth century, the department could produce thinnings, but it had to start its own tiny sawmill to process the wood because the existing sawmills and timber merchants were wedded to imported timber and were unwilling to put such a new timber as radiata pine on the market. As more logs became available in 1914, the department built a larger sawmill.[50] The other States started plantations but on a smaller scale. The relief-work of the depression enabled the area of plantations to be increased so that by 1939 Australia had 90 000 hectares of plantations, over 40 per cent of which were in South Australia.

Australian forestry cadre

The need for forestry education and training had long been appreciated; New South Wales trained its field staff within the department, and Victoria had its School of Forestry, both of which combined formal study of botany and other sciences with practical training. The forest services were staffed by these men (they all were men) and others skilled in surveying and practical work in the forests. Some, like E. F. H. Swain in New South Wales, rose to the top of their departments and became outstanding leaders of forestry; others, like Steane in Tasmania, were appointed for their administrative ability and gained forestry knowledge by experience and study. However, the imperial model demanded that the state forests should be managed by a cadre of 'professional' foresters trained at university level. In 1927 an Australian Forestry School was established in Canberra to provide two years of forestry training for recruits who had completed two years of science in the State universities.[51] Victoria held aloof from the national scheme, continuing to train its men at Creswick, but from 1942 sending a proportion of them on for a further two years at the University of Melbourne and professional standing. New graduates from the

Canberra and Melbourne schools could take positions as foresters which had previously been available only after years of experience and private study. The ranks of the services were riven further when, in 1935, the 'professional' foresters formed the Institute of Foresters of Australia and locked non-graduates out of full membership.[52] The bitterness of the rift between foresters with and without degrees lasted into the 1970s, until the latter had retired.

The resulting cadre of professional foresters, formed by common education and shared difficulties, was a small and friendly one, though with little contact between those trained in Melbourne and those trained in Canberra. While the close ties within the cadre maximised its strength in fighting the cause of forestry against agriculture, the timber industry and other parts of the bureaucracy, they created a conformist and rightist set of beliefs and values which were to prove remarkably resistant to change.[53]

Conservation organisations

As the need to conserve Australia's natural resources became more and more apparent, a number of conservation groups arose within existing organisations, such as the Royal Societies and the Australian and New Zealand Association for the Advancement of Science (ANZAAS), or as new entities. The Australian Forest League was created, after the 1911 Interstate Forestry Conference, as a lobby and public education group. An Australian branch of The Men of the Trees was formed in the 1930s after a visit of its idiosyncratic but inspiring founder, Richard St Barbe Baker, for much the same purpose; it was re-established in Western Australia in 1979 after a second visit.

Such groups were often concerned with soil conservation and catchment protection as well as forest conservation. Over-cropping, particularly during the depression, had led to dryland salinity, serious wind erosion, and sand drifts across roads, railways, and irrigation channels during the 1930s.[54] And after the 1939 bushfires Judge Stretton had seen that 'where the fire was most intense the soil was burnt and destroyed to such a depth that it may be many years before it shall have been restored by the slow chemistry of Nature'.[55] To these concerns were added those of engineers worried about the danger of silting in the new dams built for hydro-electricity and irrigation. An amalgam of these concerns led to a 'Save the Forests' campaign being launched in Victoria in 1944 to prevent bushfires, arouse public interest, and 'ensure that the water, timber and soil resources of the State shall be fully conserved'.[56] The campaign was initiated by a State politician, Cyril Isaac, and strongly supported by APM's managing director, Sir Herbert Gepp. Its council was made up of organisations ranging from the Gould League of Bird Lovers to the Hardwood Millers' Association. Forestry became part of the school curriculum and tree planting an enduring activity under the campaign's new name of the Natural Resources Conservation League of Victoria.

National Parks

Hiking became a much more popular pastime from the 1920s and so many walking clubs were formed in Victoria that they united to form a federation in 1934. Such recreation and the desire to promote tourism led to more national parks being declared. Tasmania, for example, created the Cradle Mountain–Lake St Clair National Park of 64 000 hectares in 1922.

Although most parks were far smaller, the Cradle Mountain, Wilson's Promontory (Victoria) and Royal (NSW) National Parks had established the idea that parks could be as large as 75 000 hectares. But the creation of the Kosciusko National Park in 1944 not only enlarged the size that might be conceived for a park – it is now 675 000 hectares: it changed the concept. Whereas in the political mind, parks had been simply for the delight and recreation of the people, more austere purposes were recognised. Several strands were interwoven in the change: one was the earnest, nonconformist, plain living and high thinking strand from Britain with the fell-walking and National Trust movements there; another was the wilderness strand in the United States epitomised in the writing of Aldo Leopold; another was the nineteenth-century romantic aesthetic in poetry and painting; another a transcendental attitude to nature.[57] Their Australian expression was markedly muscular. For example, the Mountain Trails Club formed in 1914 by Myles Dunphy and others enunciated a 'doctrine [of] manly and wholesome recreation' for those with 'sufficient primitive instinct [to] reach and enjoy the canyons, ranges and tops of the wildest parts'.[58] In 1933 Dunphy was instrumental in forming the National Parks and Primitive Areas Council in New South Wales, which identified the major wilderness areas of the State. He was active in having Kosciusko State Park declared and in the southern Snowy–Indy section declared as special primitive area, free from roads, tracks and huts. There was scientific interest too, as the Royal Zoological Society and the Australian Academy of Science pressed for parts of the park to be kept for scientific research, free from even Dunphy's 'people of the little tents', much to the latter's disgust. Although the Kosciusko Park was unique, it established the notion of reserving land as a wilderness accessible to none except a hardy or scientific few. It was a notion which was to become part of the environmental conflicts a couple of decades later.

Conservation and plantations in the industrial era

In the 1950s and 1960s the forest services were at last able to tackle some of the problems which their parsimonious funds had precluded before, though the burgeoning demand for wood brought new problems too. The most visible impact on the forests, apart from the greater areas being cut over, was the penetration of logging roads and fire trails into even the most isolated areas. Fire crews became much better equipped and had some hope of dealing with fires in remote spots before they 'blew up' into major conflagrations, as they had in 1939. The forests became much better known too, as aerial photography enabled foresters to make detailed maps of the forests and extend their resource assessments. Research could be afforded and was directed to the long-standing problem of regenerating cut-over areas; it was solved with a technique for sowing seed from the air.[59]

With some of the most urgent problems being dealt with, the foresters could start to prepare long-term management plans. When they had suitable data, they could calculate the 'sustained yield' of a forest; that is the non-reducible quantity of wood which could be cut from it year by year into the future, allowing for whatever variations were needed to enable the forest to reach its maximum potential rate of growth or productive capacity eventually. Their problem was to regulate the actual quantity being cut for the sawmills to the theoretically calculated 'allowable cut'. They gathered the various other forest functions under a vague concept of 'multiple use', which acknowledged that plans for managing the forests to produce wood had to be adjusted sometimes for water production, grazing, recreation, scenic and other values. In theory, 'sustained yield, multiple use' sought to maximise the net benefits to society from the combined uses.

The gulf between theory and practice became inescapably clear as it was realised that sawlogs were being cut too fast: the forests could not sustain this rate in the long term, let alone meet rising demands. It appeared that the widening gap between forecasts of demand and domestic supply could only be met by imports, unless supplies from plantations could be increased. To consider these and other problems, the Australian Forestry Council of Commonwealth and State Ministers for forests was formed in 1964, with a Standing Committee of the heads of the State forest services, and Commonwealth and CSIRO officials to advise it. It advocated more pine plantations with a will; the rate of 16 000 hectares at which the States were planting should be increased to 28 000 hectares a year, so that Australia would have an estate of 1.2 million hectares by the year 2000. In 1966 the Commonwealth provided soft loans to the States (under the Softwood Forestry Agreements) so that their planting could be increased to 26 000 hectares a year, in addition to which private growers were expected to plant 4000 hectares a year.[60] The scheme succeeded in increasing the rate of planting, and was extended, with a minor environmental qualification, for a further five years in 1971.[61]

Private plantations had been developed in the 1930s, mostly by schemes which sold area-based bonds to often gullible investors. However, the major expansion of private planting only started in 1950 when large industrial companies, wealthy enough to fund such long-term investments, commenced planting. APM planted 40 000 hectares in order to supply more pine and eucalypt pulpwood for the company's

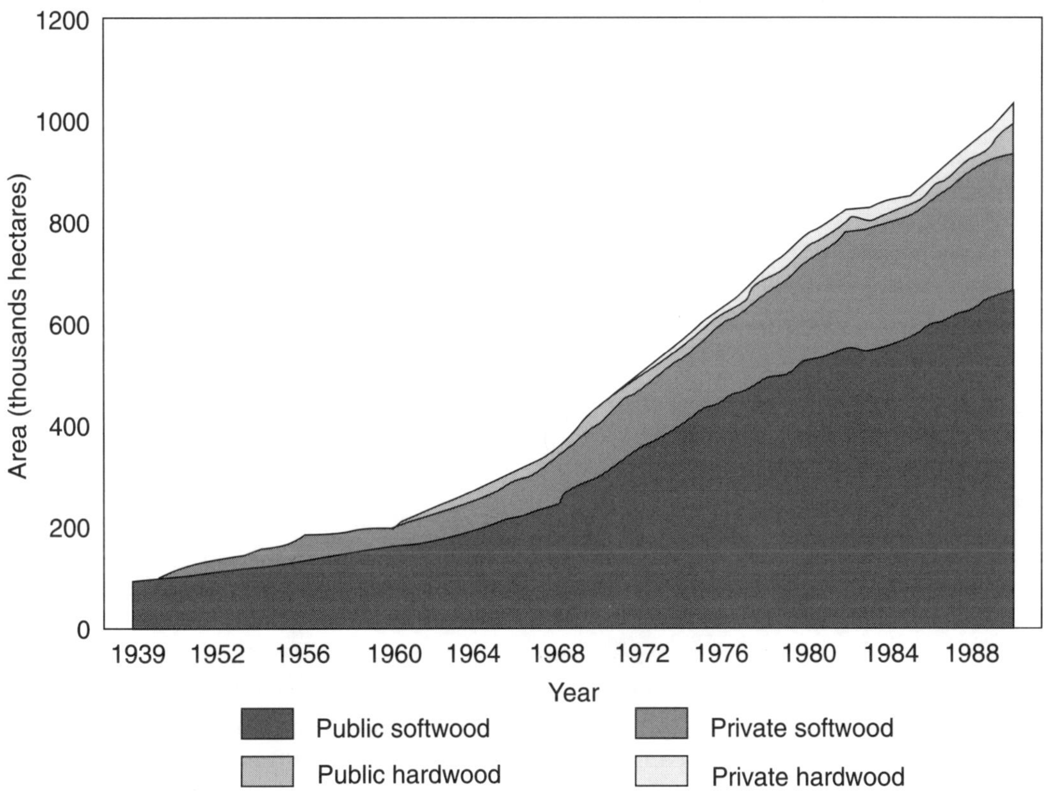

Figure 3.2 Area of plantations, 1940–90
Source: Wilson (1969); *Australian Forest Resources.*

Maryvale mill than it could get from the Forests Commission.[62] It also produced wood for a particle-board mill (1961–78) and, when the trees were big enough in 1976, for its own sawmill.[63] APPM, ANM, and a few of the larger sawmilling companies also started their own plantations to supplement resources from the state. Farmers continued to be encouraged to plant portions of their paddocks with trees, but the total area of plantations owned by farmers and a few individual investors amounts to only about 6 per cent of the private area, or 2–3 per cent of the total.

The forest services developed major programs to build roads and establish plantations. These required heavy investments which could only be provided by a change in the way they were funded. The services had been run on the basis that their expenditures should be more or less covered by their revenues from the sale of wood, with some special funds being provided during the depression and the period of reconstruction immediately after World War II. An expansion was undertaken on a large scale, it drove spending further and further ahead of income from sales of wood (Figure 3.3), and had to be funded by loans. It was claimed that the long-term plantation investments would be profitable, and hence that the debt burden would be lifted eventually when the crops were harvested.

Forest services and industry

The relationship between the forest services and industry became markedly closer. The change can be traced to World War II when foresters and sawmillers were thrown together at home to produce timber for the war effort, and abroad in Australia's 'Forestry Units' which took portable sawmills to cut timber in the war zones. Afterwards, in better economic times, perhaps the sawmillers could afford to be more accommodating, the better-staffed forest services were stronger, the endemic cheating on royalties had been overcome, and there was a greater realisation that common problems would be best addressed together.

E. F. H. Swain, with memories of the 'miserable way of life and abject poverty' of forest sawmillers and their workers during the depression, had long argued this. In 1943 he succeeded in convening meetings which became the Eastern States – and from 1960 the Australian – Timber Industry Stabilisation Conference (AusTIS).[64] Annual or biennial conferences aimed 'to cement the closest relationship between Forest Services and industry so that they may better perform their respective functions'. The methods for allocating sawmilling licences, setting royalty rates, securing 'the ultimate reconciliation of mill

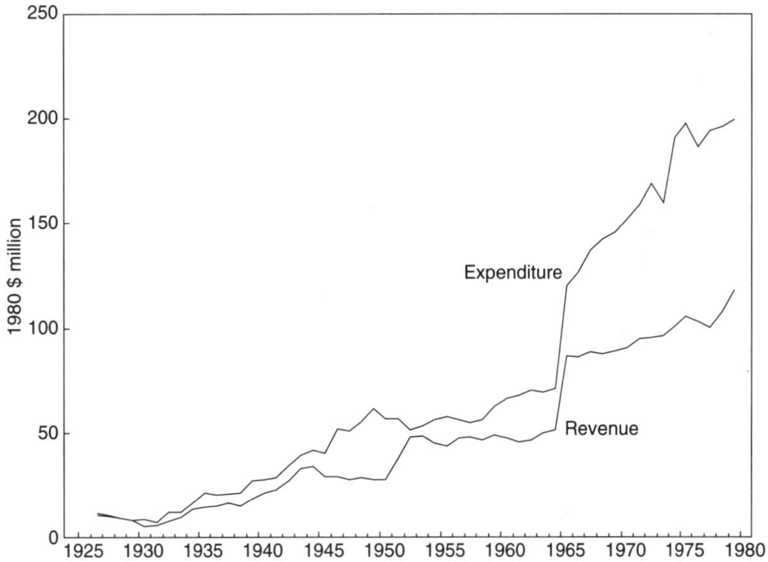

Figure 3.3 Revenue and expenditure of forest services, 1930–80
Source: Forest service *Annual Reports* for each State.

capacity with forest growth', prices and tariffs were the bread-and-butter issues, with a variety of other topics being added at later meetings. While discussions in AusTIS did evidence a shift from the pre-war days of confrontation, the forest services still had the mixed functions: they were the supervisors, tax collectors and facilitators of immediate production, as well as the guardians of the forests for future production. The foresters kept a distinct line between themselves and the sawmillers, and a greater distance from pulp and paper companies, often resenting their lush concessions and ability to deal direct with Premiers and Ministers over their heads.

The forests of the 1960s

In the 1960s the small, close-knit cadre of foresters was managing forests which were obviously varied and complex. The forests had shaped a history of use under several regimes, the impact of which had been slowly modified as forest conservation gained hold. The foresters were active in protecting the forests from clearing and fire, regenerating areas cut over, establishing plantations, and caring for the growing crops. They held a paradigm of forestry which gave primacy to sustained timber production, and believed that the energetic development of the productive capacity of the forests, supporting regional industries, would be of great benefit to Australia. The existing regimes and the very paradigm of forestry itself were soon to be tested by the three substantial changes described in the next part of this book.

CHANGING

1970s TO 1980s

This economic utilisation of non-millable material provides us with an opportunity to increase our future production by the rehabilitation and re-establishment of major forest areas ... in terms of both volume and quality ... these benefits are examples of the productivity increases which we must continually seek in Australian forestry.

Sir Reginald Swartz, Minister for National Development, at the official opening of APPM's woodchip mill in Tasmania, 1972

Conservationists regard this planet, with its rich variety of plants and animals and the only known habitat for life in the universe, as a very precious place. We consider that our human position of intellectual and technological pre-eminence gives us not a right to destroy the planet and other forms of life on it but a responsibility to protect it and them. We have an obligation to act as trustees for future generations and for other species, and to endeavour to reconcile our interests with theirs in a reasonable way.

Hal Wooten (1987), Conservation objectives, in J. Dargavel and G. Sheldon, *Prospects for Australian hardwood forests.*

The sudden advent of the woodchip export industry in the 1970s seemed to offer not only economic development, but a chance to refashion Australia's native forests into a more productive state. The Minister for Development could present it almost in the sense of conservation as wise use. But the world-wide rise of environmentalism was transforming the meaning of conservation to one of trusteeship which led the Australian Conservation Foundation, for example, to take a very different view of how the native forests should be fashioned. Indeed, much of the environmental movement thought that they should be fashioned only by natural forces; that wood production should end. In addition to these changes, the forest sector, like Australia's economy generally, was substantially restructured as jobs were shed and small mills closed with substantial regional effects.

The following three chapters examine each of these major changes in detail because it was the complexity of the increasing contradictions between them which made the environmental and other contests of the period, described subsequently, so intractable.

CHAPTER FOUR

FORESTS FOR JAPAN

During the 1970s and 1980s, economic growth slowed or even declined in many countries, industries were restructured, unemployment rose sharply, and the numbers of the world's poor increased. In contrast, two groups of countries did well. The oil-rich, mostly Arab, OPEC countries greatly benefited from higher oil prices which, in the early 1970s, they largely engineered. The other group in East Asia comprised Japan which was already industrialised, and the 'newly industrialising countries' (NICs) of Hong Kong, Singapore, South Korea and Taiwan; together, they became a second generation of the world's affluent countries.[1]

Japan was outstanding. During the 1950s, it managed to reconstruct its cities and industries and achieve an average rate of economic growth of over 9 per cent a year, only to exceed this by the 'economic miracle' of a 10 per cent growth rate during the 1960s. Although the rate slowed during the 1970s and 1980s, the relative wealth of its citizens continued to increase. In 1960, Japanese people had a per capita income less than one-third of that of Australians, but by 1985 they had passed Australians and by 1988 the North Americans; they had the highest incomes in the world.[2] Japan's economic rise came from its success as a manufacturing and trading nation. As the limits of its various natural resources were reached, it increased its imports of raw materials.[3] These provided Australia with a welcome offset to the general economic malaise. Australia was rich in minerals and energy resources. Exploration flourished, new mines were opened, railways and towns built in remote parts in a new 'resources boom'. Australia also had forest resources. How these too were rapidly swept up into large-scale projects to export woodchips to Japan is the subject of this chapter. To understand why this happened so suddenly, it is necessary to start the story in the forests of Japan.

Japanese markets and forests

Japanese forests still cover two-thirds of that mountainous country and have for centuries provided timber, wood, delicious shiitake mushrooms and a host of other forest products. In pre-industrial Japan, periods of increasing exploitation had led to regional shortages, erosion and silting, which had been countered by conservation measures.[4] Well before the first convict ships had sailed for Australia, the demand for timber had pushed prices up enough for farmers near the cities to start growing high-quality sugi and hinoki logs in plantations. In the nineteenth century, Japan's industrial development relied heavily on its forests for energy as well as for construction. Charcoal and fuelwood supplied as much as 85 per cent of Japan's energy in the 1880s, and charcoal continued to cook food and warm homes right into the 1960s.

Towards the end of the nineteenth century, national and imperial forests were declared, a forest service established, protection forests reserved, and planting accelerated to a peak of 150 000 hectares a year in 1910.[5] Less than one-third of Japan's forest area was included in the national forests; much of the rest was in very small parcels owned by a largely mountain populace. Even in 1980, 2.5 million households still owned forests, and of these 89 per cent had 5 hectares or less. The mountain communities were substantially self-sufficient, but they relied on growing and selling wood, charcoal and silk for cash income. Their traditional lifestyle contrasted sharply with the modern lifestyle emerging in the industrialising cities.

Forest resources were also obtained by imperial expansion, particularly from the southern half of Sakhalin Island which Japan took from Russia in 1905. By the 1920s, its forests were providing 69 per cent of Japan's pulpwood, part being used in eight local pulp mills and part being shipped to Honshu.[6] A little also came from Korea, which Japan had conquered by 1910. Otherwise, forest production, based on low labour costs, expanded more or less to meet demand, so that timber imports and exports were of comparatively minor importance.[7]

The Japanese cut their forests hard during World War II and for reconstruction afterwards. By the mid-1950s, 2 million hectares needed replanting and the growing stock was badly depleted; the demand for wood was double that before the war, and was expected to double again.[8] Reforestation was undertaken energetically by planting 300 000 to 400 000 hectares a year during the 1950s and 1960s. Although rates reduced in the 1970s and 1980s, 10 million hectares, which amounted to 40 per cent of Japan's forest area, were reforested. In the process, large areas of hardwood and mixed forest were converted to conifer plantations, partly because sugi, hinoki and other softwoods were preferred in construction, but also because the rising pulpwood market was primarily for softwood, while the collapsing firewood and charcoal markets were for hardwood.[9]

In the post-war era, young people left the mountains for the cities, farming for subsistence was replaced by farming for the market, traditional leadership structures weakened, and community services deteriorated.[10] Steep slopes, remote areas, poor infrastructure, small holdings and much higher labour costs increased the costs of producing wood, even when land-holders continued to grow it. Thus, Japan's ability to supply its rising demands for forest products during the 1960s was limited by the loss of its colonies, the past depletion of its forests, the immaturity of its extensive new plantations, and the ending of a rural structure on which so much of its production had depended. The total volume of timber cut declined from almost 80 million cubic metres in 1961 to only half that in the twenty years that followed. This was accompanied by rising consumption in the cities: imports were the only solution, the forests of the Pacific Basin their source.

Japanese pulp and paper industry

The sudden rise of Japanese domestic affluence is epitomised by the consumption of paper and paperboard. In 1960, the Japanese on average consumed 46 kilograms per person each year, while the Australians consumed double that; but by 1990 the Japanese were consuming 228 kilograms a year, well behind the Americans (311 kg), but well ahead of the Australians (165 kg). The population also increased, so that Japan's total consumption increased by nearly six times.

Japan was able to meet its soaring demand for paper by expanding its pulp and paper industry, rather than by importing paper.[11] This was aided by the location of many of the paper companies within Japan's industrial conglomerate groups, the *keiretsu*. These contain diverse companies whose activities are coordinated by an overall presidential council and cross-holdings of shares. Finance is arranged through particular banks, while trade is conducted through the giant multinational trading companies, the *sogo shosha*. These buy raw materials around the world, organising their production if necessary; they sell goods for Japanese manufacturers; they arrange and often own shipping; they finance trade and sometimes venture their own capital in manufacturing; and they maintain their own communication

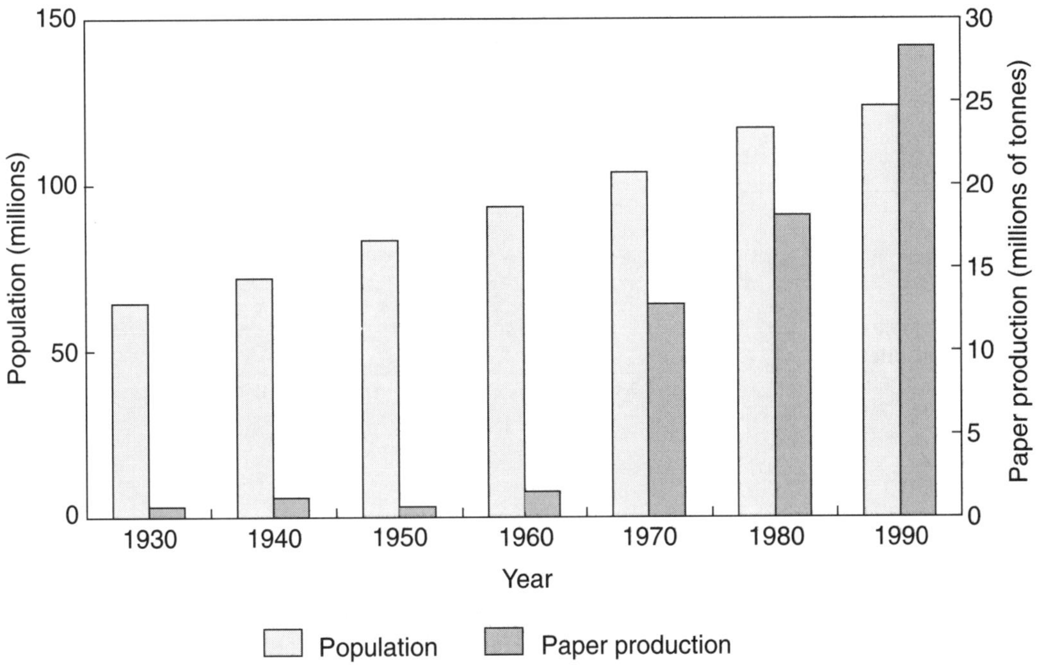

Figure 4.1 Population and paper production in Japan, 1930–90
Source: Japan Statistical Yearbook 1991; Handa 1989.

networks linking their offices around the world.[12] In the pulp and paper sector, they trade between North American and European exporters and buyers throughout East Asia, which gives them an unrivalled knowledge of market trends. Within Japan, the keiretsu link them with several of the major pulp and paper companies, while the Japanese Ministry of Trade and Industry (MITI) exercises a powerful guiding function.

Such rapid expansion depended on finding new sources of raw material. As the domestic supplies of softwood pulpwood from the mountain forests diminished in the 1960s, the industry turned to hardwood pulpwood and to chips produced mostly from the waste at Japanese sawmills. It also started to import small quantities of pulplogs from the Soviet Far East, mostly from Sakhalin. However, loading logs on and off ships is an awkward, expensive business which limited foreign trade until a method for bulk handling was developed in 1965. This involved constructing woodchip mills that cut the large pulplogs

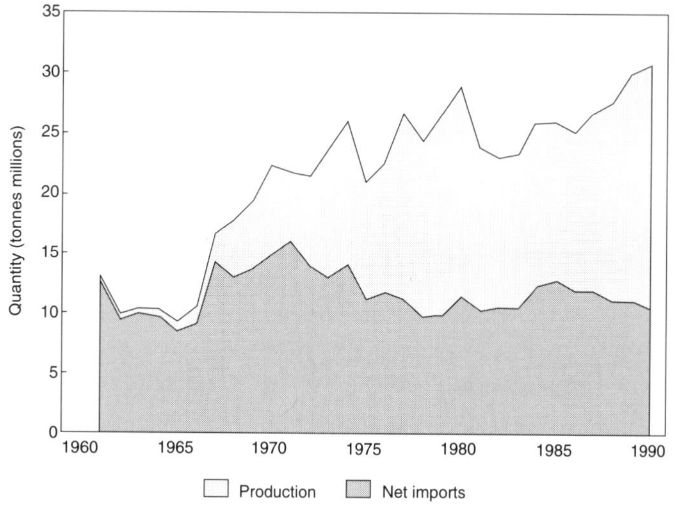

Figure 4.2 Japan, pulpwood production and imports, 1960–90
Source: Handa 1989; FAO *Yearbook of Forests Products* 1990.

into chips able to be handled in bulk by bulldozers and conveyors, or blown along chutes. Importantly, specially-built bulk carriers could transport woodchips cheaply, which brought distant Pacific shores within economic radius.

Finding 10–15 million tonnes of woodchips and bringing them across the oceans each year would have been a formidable task for the paper manufacturers were it not for the sogo shosha. They sent men around the Pacific to search for wood. They found softwoods in the United States that were easy to pulp and which they bought readily. From initial shipments in 1965, they built up the trade to over 7 million cubic metres a year in a decade. Hardwoods were more difficult. In the tropics, logging is hard and the jumble of different woods makes it almost impossible to make a uniform pulp. However, further south, in Australia, they found the eucalypt forests. Australia had the infrastructure and stable government of a developed country, the experience of forty years of research which showed how fine papers could be made from the wood, and an established logging industry which would produce it. They started an Australian trade in 1970 and in four years built it up to 2.6 million tonnes a year, since when it has doubled.

Besides from importing woodchips, raw materials for making paper were obtained by importing ready-made woodpulp and by recycling paper waste. Supplies from both sources were increased after the OPEC oil crisis in the mid-1970s made them relatively more attractive. Pulp imports increased from about 5 per cent of the world's trade in 1965 to 14 per cent, or 2.6 million tonnes a year, by 1990. However, it was increased recycling that had the greatest effect. Gradually, Japan managed to raise the proportion of the year's production that it recovered as waste for use again to reach 50 per cent by the mid-1980s, one of the highest rates in a developed country.[13] It aims to reach 55 per cent by 1995. Australia, which did only half as well as Japan in the mid-1980s, increased its recovery rate too and expects to reach 43 per cent in 1995.[14] Japanese paper companies put the wastepaper into about 25 per cent of the furnish for making paper, and into a remarkable 85 per cent of the furnish for making paperboard. By 1990, 52 per cent of the pulp being fed onto Japan's paper machines was coming from their own waste and 18 per cent was coming as new pulp made from Japan's own wood; this left 30 per cent to be imported – 9 per cent as pulp and 21 per cent as woodchips which were made into pulp in Japanese mills. Australia supplied about one-fifth of the woodchips and hence only about 5 per cent of the fibre used.[15] That such a small proportion could create such conflict in Australia is a measure of the impacts of Japan's burgeoning consumerism on the Pacific Basin.

Australian woodchip trade

No one in Australia expected serious conflicts about exporting woodchips. Quite the reverse: a new export staple seemed a welcome addition to the long history of wool, gold, wheat, meat and all the other exports that had driven Australia's development. And the energetic pursuit of development was still very much part of Australia's ethos. Moreover, there were particular attractions to woodchip exports: they would bring some development and employment to backward rural regions, they could provide a first stage which could lead to pulp and paper manufacturing as the Australian market grew, they could transform the forests to a more productive state, and they might be highly profitable to the exporters. The Commonwealth's Forestry and Timber Bureau could even assure the Australian Conservation Foundation that 'the forest activities are confidently expected by foresters to result in an improvement in the health of the forest and an increase in fauna'.[16]

A euphoric atmosphere pervaded business and State governments as team after team of the sogo shosha and pulp company men arrived from a bewildering array of unfamiliar companies. The linkages between them were little realised; nor, at first, was the degree to which their negotiations over resources and prices, and their joint investments in shipping, were being orchestrated by MITI's Overseas Pulp Materials Committee. Government departments and Australian firms tumbled over each other in an almost farcical manner to find resources and make frantic proposals to export them. It all happened very quickly. Serious negotiations started in 1966, a mill was operating in New South Wales in 1970, and its first load of woodchips sailed out in January of the following year. By 1972, three more woodchip mills were exporting from Tasmania, and by 1976, the last of Australia's five big mills was exporting from Western Australia. The 1 million tonnes of pulpwood that were being cut for the Australian paper mills had been increased to 4 million by the Japanese trade. In the following years, a few smaller producers were added and the big mills increased their output so that by 1990, Australia was exporting 5 million tonnes, most of it to Japan. The pulp mills in Taiwan and South Korea also needed to import wood,

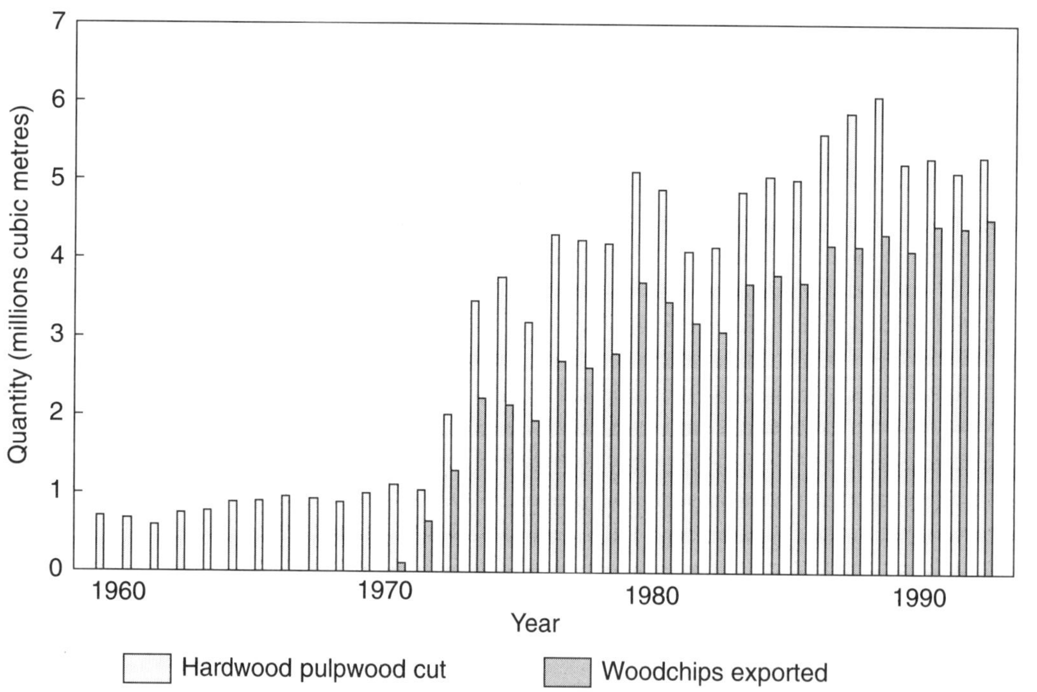

Figure 4.3 Woodchip exports and total pulpwood cut from Australian forests, 1960–93. Woodchip exports include chips produced from sawmill waste and pine plantations.
Source: Australian Forest Resources, various years.

but on nowhere near the scale of the Japanese mills. A small trade was started, which is not further discussed because it amounted to only about 2 per cent of the Japanese trade by the 1990s.

In the flurry to get into the trade, Australian firms seemed ready to sell the resources for almost any price they could get. Something better was needed if Australia as a whole was to benefit, and in September 1968, the Commonwealth decided to include woodchips among the commodities over which it exerted its constitutional power to control foreign trade. The controls it introduced required that: the contract

Table 4.1 Industrial structure of woodchip export trade, 1994

Australian exporter	*Mill location*	*Quantity licensed (000 tonnes)*	*Trading house*	*Imported into*	*Pulp and paper company*	*Notes*
Hardwood						
North Broken Hill Peko						
North Forests (APPM)	Tamar, Tas.	1065	Mitsubishi Sumitomo	Japan	Mitsubishi Paper Nippon Paper Shin Oji Paper	
TPFH	Triabunna, Tas.	813	Mitsui	Japan	Nippon Paper	
Gunns	Tamar, Tas.	200				Also from Smithton, Tas.
Boral						
Forest Resources	Tamar, Tas.	947	Kawasho KDI CS Lui	Japan Korea Taiwan	Daio Paper Chuetsu Dong Hae Chung Hwa	Kanematsu buys spot sales not required by Kawasho
Sawmillers Exports	Newcastle, NSW	500	Itochu Toyomenka	Japan	Chuetsu Shin Oji	Tomen buys some spot sales
Bunnings WA Chip & Pulp	Manjimup, WA	970	Marumbeni	Japan	Nippon Paper Hokuetsu	Incl. 70 000 tonnes from plantations
Harris–Daishowa	Eden, NSW	900	Itochu	Japan	Daishowa Paper	Incl. 20 000 tonnes from East Gippsland, Victoria
Brisbane Woodchips	Brisbane, Qld	180	Itochu	Japan Taiwan	Shin Oji Chung Hwa	
Midway Forest Products	Geelong, Vic.	170	Mitsui	Japan	Nippon Paper	
Softwood						
SEAS Sapfor	Portland, Vic.	700	Mitsui	Japan	Nippon Paper	
QCE		400	Sumitomo			
North Broken Hill Peko	Burnie, Tas.	300				

price for sales be at or above the ruling world market price, after making allowances for differences in quality and transport costs; that they be adjusted for inflation periodically; and that further processing in Australia be investigated and started if it proved feasible. The Commonwealth added further requirements on other matters where that State provisions seemed inadequate.[17] The State governments provided concessions over the forest resources under existing or new legislation. Thus, the various export projects, described in turn below, were legitimated by both Commonwealth and State governments.

Eden woodchip mill and forest concession

Woodchip exports not only happened quickly, they happened on a large scale in projects which required substantial forest resources and deepwater anchorages for loading the ships. The south-east of New South Wales readily met the requirements: Eden had a sheltered harbour in Twofold Bay and a large area of public forest behind with suitable resources. Indeed, these forests had first been investigated for a domestic pulp and paper mill in the 1930s.[18]

Harris–Daishowa's woodchip mill at Eden, NSW. (Australian Conservation Foundation; National Library of Australia)

In the 1960s, Eden was dependent on farming, fishing and tourism, but the high cost of transporting its produce nearly 500 kilometres to Sydney meant that the region was economically peripheral. Its forest industry was limited to a few hardwood sawmills, only one of which was of any size. In this situation, the Forestry Commission sold sawlogs from the public forests at the token royalty rate of only about 50 cents per cubic metre, which raised barely enough revenue to pay for a minimal level of supervision, fire prevention and administration.[19]

The rather dry coastal forests behind Eden and south to the Victorian border were the most accessible. Silvertop ash was the main species, mixed with messmate, white and other stringybarks, and with big grey gums in the gullies. These forests had been logged, in some places more than once, but only trees with good sawlogs had been taken. Severe bushfires had ravaged the region periodically so that most of the remaining trees had little value for sawlogs, although there was plenty of good pulpwood in them. Luckily, the forest easily regenerated naturally after fire. The roads through the forest were very lightly built, most having been put in by the sawmillers for use in dry weather, and some were so poorly located that they ran through streams. Behind the coastal forests were moister ones which ran back up the escarpment to the tablelands. These higher, wetter areas had some magnificent stands of messmate, brown barrel and shining gum, with tall, straight, large trees containing both fine sawlogs and pulpwood. Many of the areas had not yet been reached by logging, and it was apparent that some major roadworks would be needed. There were some national parks in the area, and the large Nadgee nature reserve on the coast which it seemed that hardly anyone ever visited.

The New South Wales government was keen to turn the natural resources of the backward Eden region to good effect, and called for bids as soon as it learnt of the Japanese market. Apart from having a decent map, the Forestry Commission knew virtually nothing about the pulpwood resource; it did not even have a volume table from which to calculate the quantity of pulpwood in the district's trees. Nevertheless, its officers pored over aerial photographs, measured plots of trees, and made some rough estimates of the resource. There was obviously enough pulpwood for twenty years, and probably forty, in the forests lying south of Bega between Bombala and the coast. Better estimates were made, but only well after the woodchip mill had started.[20]

Victoria had similar forests in East Gippsland which its Forests Commission assessed carefully. However, the East Gippsland coastline to stormy Bass Strait lacked anywhere sheltered enough for ships to be loaded with woodchips. Although some of the forests were within economic radius of Eden, Victorian politicians wanted industrial

Figure 4.4 South-east Australia, showing the Harris–Daishowa woodchip concession

development in their State and were loath to see wood trucked over the border. The Commonwealth seems to have made no attempt to overcome State rivalries and mediate a larger project, either at the time or since.

Several companies bid for the New South Wales pulpwood, including APM with its blue-ribbon status and long experience. Another was the Harris–Daishowa consortium put together by a Sydney entrepreneur, trading as Harris Holdings Ltd, aided by one of the smaller local sawmillers, E. B. Cane, who had once worked for the Forestry Commission. It declared it had a 51 per cent Australian ownership in the consortium, while the rising Japanese paper company, Daishowa, held the other 49 per cent. It negotiated the murky politics of the Askin government, and in 1968 signed an agreement with the Forestry Commission which allowed 508 000 tonnes of pulpwood to be cut from public forests within 80 kilometres radius of Eden each year for twenty years. The Premier announced that the company had undertaken to build a pulp mill in due course. Wood was also available from private land in the region and as chips produced from sawmill waste. To cover all sources, the Commonwealth issued an export licence for 636 000 tonnes for the same period. The Commonwealth clearly regarded woodchip exports as only a preliminary stage, because its approval was conditional on the company starting to build a pulp mill by 1977. Harris Holdings could not finance its share of building the woodchip mill and was replaced briefly by other Australian companies, before the project was taken over by Daishowa (62.5 per cent) and C. Itoh (37.5 per cent) who arranged the shipping.[21] Only the name, Harris–Daishowa, and one Australian director, E. B. Cane, were retained.

Itoh built the world's largest woodchip carrier, the *Empress of Eden*, and Daishowa built the woodchip mill. The main difficulties occurred in the forest, partly because the roads were quite inadequate for large-scale logging operating all the year. The Minister hurried the project so that New South Wales should start before Tasmania, but the slow-moving Forestry Commission failed to plan, budget, staff or manage the task of starting such a large new project in time, and the company failed to recruit experienced logging managers. The logging contractors had to cut in the few places they could get to, and for the first few years, these were all in the coastal forests near to the mill where earlier sawmill logging had left at least some roads. The result was a hideous mess that spread across the landscape near the Princes Highway.

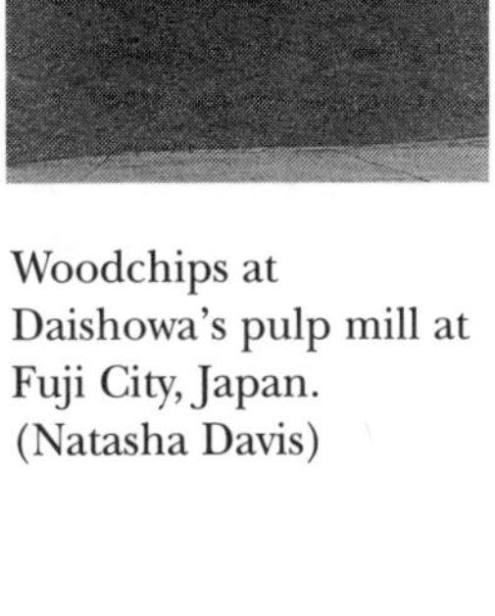

Woodchips at Daishowa's pulp mill at Fuji City, Japan. (Natasha Davis)

The environmental impacts, objections and conflicts that resulted are described in subsequent chapters. It is sufficient here to note that the Forestry Commission belatedly came to provide a competent planning and management organisation, and built a network of substantial roads which enabled the logging to be dispersed across the concession area. Harris–Daishowa hired experienced logging managers, improved the mill and was able to raise its output.

As Daishowa increased its paper production in Japan, Harris–Daishowa had to find more wood and get its export licences increased. In 1975, it entered into a further agreement with the Forestry Commission, which included wood from private property, waste from sawlog operations outside the concession boundary, and sawmill residues. With this in hand, it was able to get the Commonwealth to increase its export permit to 762 000 tonnes a year. But this was still not enough, and three years later, it proposed raising production to 850 000 tonnes a year by 1978. This time the Commonwealth ordered an Environmental Impact Statement to be prepared, but allowed an increase to 806 000 tonnes while it was done. Once it was, the quantity was increased to 850 000 tonnes for the period to 1989. Export approvals were also granted in 1981 for an extra 50 000 tonnes of wood salvaged after a bushfire; and in 1982 for an extra 100 000 tonnes a year for five years from trial operations in Victoria. By 1985 the environmental conflicts over continuing the export of woodchips had increased, and a further Environmental Impact Statement was called for.

Through all these increases, the Commonwealth and the New South Wales governments still hoped that Daishowa would turn the woodchip mill into a pulp mill, thus adding value and employment in the Eden region. Although it called for Daishowa to produce feasibility studies, it could not force it to build a pulp mill, nor in practice could it cancel the licence if it did not; there were too many already employed there. Daishowa had other plans; it upgraded and expanded its five pulp and paper mills in Japan. Two of Daishowa's mills depended on imported chips: the Suzukawa mill in Fuji city was almost totally dependent; and the company's largest mill, the Shiraoi mill in Hokkaido, which used almost 1.5 million tonnes of wood a year, had to import half of it as chips.[22] Rather than lose wood resources for its existing pulp mills, it expanded its paper-making in Japan with imported pulp and recycled paper. It became the most venturesome of the Japanese pulp and paper companies as it bought two export pulp mills in British Columbia (one as joint venture with Marubeni, a sogo shosha), and built a very large pulp mill of its own at Peace River in Alberta. Each produces bleached kraft pulp.[23]

It was not for Eden that Daishowa put out a proposal in 1989 to build a pulp mill, but for the Clarence River in northern New South Wales. It seemed an unlikely venture, since the company was buying and building pulp mills in Canada at the time. However, the prospect fitted neatly with expansionary proposals being floated by the Forestry and Forest Products Industry Council and an industry task force in New South Wales.[24] Real or not, the proposal was used to justify renewing the Eden woodchip concession without a pulp mill; perhaps the government simply realised that there was not much else it could do. Although the environmental conflict over continued logging in the Eden forests was sharp and the political process which followed the Environmental Impact Assessment was convoluted, New South Wales renewed the forest agreement and the Commonwealth renewed its approval so that 850 000 tonnes of woodchips a year could be exported until 2004. Another 150 000 tonnes a year from Victoria was added in 1989.

Triabunna woodchip mill and forest concession

The prospects for woodchip exports from Tasmania were bright, because both forest resources and sheltered loading points were plentiful. The first of the three mills was started by a group of thirteen sawmillers who combined to form Tasmanian Pulp and Forest Holdings Pty Ltd (TPFH).[25] The member companies included long-established Tasmanian timber firms such as Risby Bros, Kilndried Industries owned largely by a mainland timber network, and Tasmanian Board Mills which had attempted to enter the pulp and paper industry in the 1950s.[26] They had substantial logging experience, were well known in the parochial atmosphere of Tasmanian politics, and together could raise Australian capital. The woodchip industry offered them the prospect of breaking into the pulp and paper industry, hitherto the preserve of big companies.

The resource was secured when the State government granted a concession over uncommitted Crown forests in eastern and southern Tasmania which provided exclusive rights to 508 000 tonnes a year of pulpwood for eighteen years. Australian Newsprint Mills (ANM) objected, claiming it might need some of the wood for its own long-term expansion, but the government gave it short shrift, partly because the case was weak, and partly because it wanted to encourage a large new venture, especially one which appeared to be particularly Tasmanian. The concession agreement required TPFH to conduct a feasibility study into building a pulp mill. If one was built, the resource rights would be extended to eighty years; otherwise, exports could be continued on a basis to be negotiated. In addition to the Crown wood,

TPFH also obtained wood from privately owned forests and from chips produced from the waste in its members' sawmills. A sheltered point for loading was easily found at Triabunna on the east coast, and a market was secured by entering into a 15-year contract with Mitsui, which arranged to supply 610 000 tonnes a year to the Jujo Paper Company. The initial price of $13.50 per tonne is said to have set a record, and the Commonwealth gave its approval for 710 000 tonnes a year to be exported.

Mitsui built two bulk carriers, the *Nego Triabunna* and the *Silviculture*, which started taking the chips to Ishinomaki in northern Japan in 1971. It was harder for the Tasmanians to raise the money to build the mill, but Kilndried Industries increased its shareholding, the AMP Society provided $1.5 million, and loans of some $4 million were raised.[27] Unlike Eden, logging worked well from the start. Tasmania's Forestry Commission, although much smaller than that of New South Wales, had experience with domestic pulp mills, knew much more about its resource, and could plan the operations properly. Notably, the company spent $1.5 million a year for the first few years on building a very solid, all-weather, main logging road through the concession.

In 1973–74, TPFH wanted its permit increased by a further 204 000 tonnes so that it could export more wood, mostly from private property. The Commonwealth did not grant the application because the Tasmanian government was worried about the amount being cut from

Logs which are too bent or defective for sawlogs can be used for pulplogs. TPFH woodchip concession, Tasmania, 1979.
(John Dargavel)

private forests. The major review came when the whole State's woodchip export program was examined in an EIA, and eventually the licensed quantity was increased to 813 000 tonnes a year. But by then, TPFH had been taken over by APPM, who also built a woodchip mill.

APPM's Wesley Vale concession and Tamar woodchip mill

Associated Pulp and Paper Mills Ltd (APPM) was the most experienced and wealthiest of the Australian firms to enter the woodchip trade. It had been making pulp and paper at Burnie, in north-west Tasmania, since 1938, and had made good but fluctuating profits through the 1960s; with $10 million in hand, and $11 million raised in share and debenture issues, it could readily finance a new project.[28] Moreover, it already had an unused concession over 512 000 hectares of public forests in northern Tasmania. This had been granted in 1961 to encourage the company to build a pulp and paper mill at Wesley Vale, but it had proceeded cautiously and by 1970 had built only the paper mill; woodchip exports presented an opportunity both to turn a profit and to hold the concession.

APPM asked the government for permission to export the wood as a temporary phase, stating that it would build the pulp mill by 1978. A group of sawmillers also applied but were refused,[29] and APPM was allowed to cut 508 000 tonnes a year from the concession and a further 255 000 tonnes a year for five years from uncommitted state forests in the north-west. Further wood was available from private property and from sawmill waste. The Commonwealth approved export permits of 610 000 and 365 000 tonnes for eleven and five years respectively. The State licences and Commonwealth export approvals were revised and extended at various times. After the major environmental contest of 1984–87, it was authorised to export 1 065 000 tonnes a year until 2003.[30]

Two sogo shosha, Mitsubishi and Sumitomo, contracted with the company to supply Mitsubishi Paper Company and Sanyo–Kokusaku Paper Company. With markets, money, wood resources and substantial engineering experience, APPM proved an effective supplier. It built a rugged woodchip mill at Long Reach in Bell Bay in the quiet estuary of the Tamar River. The mill started ahead of schedule in 1972 and operated most efficiently. The concession already had many public roads through it, but to reduce the number of log trucks pounding through Launceston, the company brought part of its wood by rail.

APPM made a major expansion in 1979 by winning a vigorous takeover battle with Forest Resources for TPFH. To do so, it bought out the entire Tasmanian interests of the network of mainland timber merchants, including Kilndried Industries, sawmills and a ply mill. It became Tasmania's largest sawmiller, had licences to export nearly 2 million tonnes of woodchips annually, and held pulpwood rights over three-quarters of Tasmania's state forests. In spite of its dominant resource and production position, the 1982 recession reduced APPM's profits and it feared a hostile takeover; this led its largest and founding shareholder, the North Broken Hill (NBH) mining company, to buy out the other shareholders in 1983.[31] NBH also controlled the mines

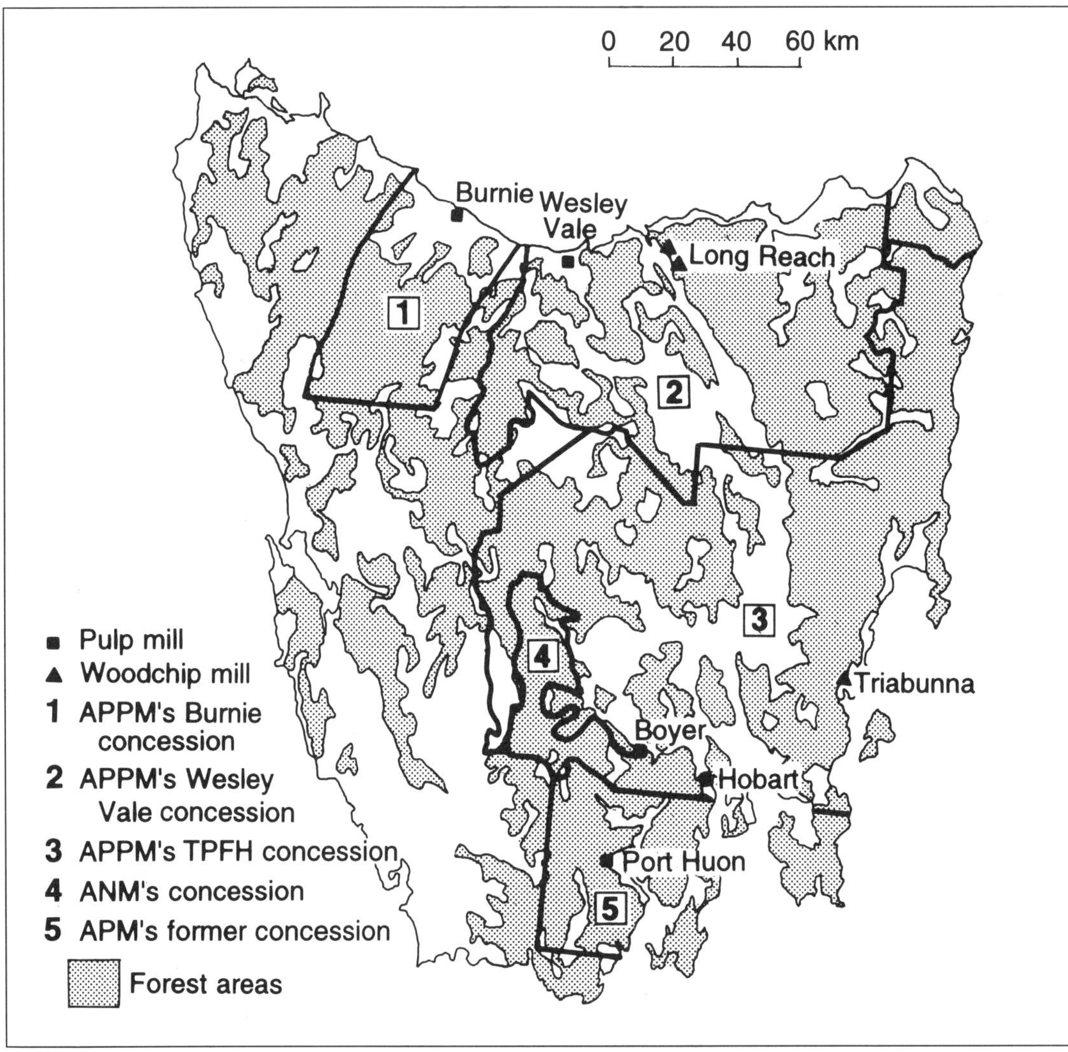

Figure 4.5 Tasmania, showing forest concessions

and zinc smelter of EZ Industries in Tasmania, which made it by far the largest employer and led to Tasmania being depicted with some hyperbole as the first 'company State'. Its clout was further strengthened by amalgamating with another mining company to create North Broken Hill Peko Ltd. In 1988 NBH made Australia's first, apparently serious proposal to upgrade a woodchip mill into an export pulp mill to be built at Wesley Vale. This raised such a storm of environmental protest and major political conflict that the proposal collapsed, and the resource continued to be shipped out as woodchips.[32]

Forest Resources' Tamar woodchip mill

The group of sawmillers, Tasmanian Woodchips Pty Ltd, who had tried to get wood from APPM's unused Wesley Vale concession were not put off when the government preferred APPM's application with its promise of a pulp mill; instead they turned to the pulpwood resources on private land and to waste from its sawmills. They obtained cutting rights on as many properties as they could and negotiated a contract to supply 305 000 to 710 000 tonnes a year to Japan. Its price of $11 per tonne was so far below the TPFH price of $13.50, that the Commonwealth refused an export permit.[33] The sawmillers then formed Northern Woodchips Pty Ltd with the Savoy Corporation Ltd, renegotiated the contract at a higher price and obtained Commonwealth approval to export 9.1 million tonnes over a 15-year period. No sooner had Northern Woodchips started to build a woodchip mill next to APPM's on the Tamar River, than the Savoy Corporation collapsed. However, the project was saved from bankruptcy and refinanced by an oil company, H. C. Sleigh Ltd, and the Commonwealth's Industry Development Corporation which took a 20 per cent equity. The contract was renegotiated with Kawatetsu Bussan (later absorbed into Kawasho) to supply the Taio Paper Company. The Commonwealth reissued its approval which allowed 710 000 tonnes a year to be exported, but added conditions which required the company to study the feasibility of building a pulp mill, and if feasible, to have one operating by 1986. It also had to reforest at least 2000 hectares a year of the forests it felled on private land.

The project encountered many difficulties. The chip mill was very lightly constructed, so that breakdowns were common and a major reconstruction was necessary. Without exclusive rights in a forest concession, the company had to range far and wide to buy wood from private properties throughout northern and central Tasmania, often in the face of hostile competition from APPM. APPM also managed to block Northern Woodchips from buying most of the waste from sawmills by claiming rights to all waste cut from any logs within its concessions.[34] Nevertheless, Northern Woodchips improved its mill, overcame the competition for resources and, by 1979, increased its output to the maximum of its export licence. In the 1980s it established a rail loading yard in the south of the State and was able to bring unwanted wood from the ANM and APM concessions there. It also started cutting smaller wood and salvaging dead wood. Such increased resources enabled the company to obtain approval to increase its exports by 103 000 tonnes a year to Japan and 90 000 tonnes a year to South Korea. Various revisions followed, so that by 1990, the company had approval to export 947 000 tonnes a year until 2003.

H. C. Sleigh provided the money and management to develop the project successfully, and in 1977 exercised an option over the government's shares, bought out the remaining shareholders, and incorporated Northern Woodchips as a division within the corporation, known as Forest Resources. Like APPM, Sleigh was swept up in the restructuring of Australian industry, becoming first Petersville Sleigh, and finally, in 1992, part of the Boral group, which owned another woodchip exporter, Sawmillers Exports, in New South Wales.

Western Australian woodchip mill and concession

The last of the big five projects was built in Western Australia to export chips from the marri and karri forests in the south-west through the port of Bunbury. The resource was made up of waste from the sawmills cutting karri, and from marri trees which were little used for sawmilling but were suitable for making paper pulp. The marri were intermingled through much of the jarrah forests and to a lesser extent in the karri forests. They had largely been left behind after the prized jarrah and karri had been felled, and presented an increasing hindrance to the regeneration of the prized species. Hence, an opportunity to fell them for woodchips was particularly attractive to the Forests Department.

Two proposals were put to the government in 1967, which finally decided to accept one prepared by the Western Australian sawmillers, Bunnings Timber Holdings Ltd. Its scheme envisaged building a woodchip mill at Manjimup in the middle of the karri belt, and taking the chips 120 kilometres by rail to Bunbury where a new port was being built. However, the project did not start until 1976, during which time there were several changes. First, Bunnings took over the Western Australian timber operations of a long-time rival, the British Hawker

Clear-cut coupe in karri-marri forest, Pemberton District, WA. Note trees left to provide seed and wildlife habitat.
(Colin Totterdell, Australian Heritage Commission)

Karri and marri regeneration growing on coupe cut two or three years previously, Pemberton District, WA.
(Colin Totterdell, Australian Heritage Commission)

Siddeley company, which doubled Bunnings' size and timber resources. Second, the original sogo shosha found other suppliers and were eventually replaced by the Marubeni Corporation, which sent the chips to three Japanese paper mills. Third, increasing environmental concerns over woodchip operations were evident in Western Australia, where there were already particular concerns over salinity. In 1973 the State government had an EIS prepared urgently and set up a Technical Committee to identify salt-sensitive areas. Fourth, a woodchip export company, Western Australian Chip and Pulp Pty Ltd, was formed, with Bunnings having the largest shareholding (36 per cent), and the other major sawmilling company in Western Australia, Millars, having a significant holding (28 per cent). The remaining capital was subscribed by the Australian Industry Development Corporation and the AMP Society.

The mill finally started with a State licence to cut 680 000 tonnes of pulpwood a year from state forests. In addition it obtained chips from the sawmills and some wood from private forests, making a total of 750 000 tonnes a year. A Commonwealth export licence was granted for fifteen years and subsequently extended to 1998. Like the Tasmanian woodchip mills, the ownership of WA Chip and Pulp changed as Australian industry restructured into larger firms. Bunnings merged with Millars in 1983 and made the woodchip company a wholly owned subsidiary, in what was almost a monopoly of the hardwood timber industry in Western Australia.

Other woodchip exports

In the heady atmosphere of the resource boom, many other attempts were made to develop woodchip exports at places spread from Arnhem Land and Cape York to southern Tasmania. However, poor wood quality in northern Australia, inadequate quantity, lack of suitable ports and rising environmental objections meant that only four further projects eventuated. They were developed later than the 'big five' and encountered more environmental opposition, so that governments only approved them on the basis that no additional clear-felling of the native forests would be required.

The forests strung out behind the central and north coasts of New South Wales were accessible to several possible ports and contained pulpwood, though not in such large quantities near a single port as the earlier projects. Three proposals were submitted in 1975, complete with their newly-required EISs.[35] What resulted was based on exporting only chips made from sawmill waste, or from the logging residues of butts

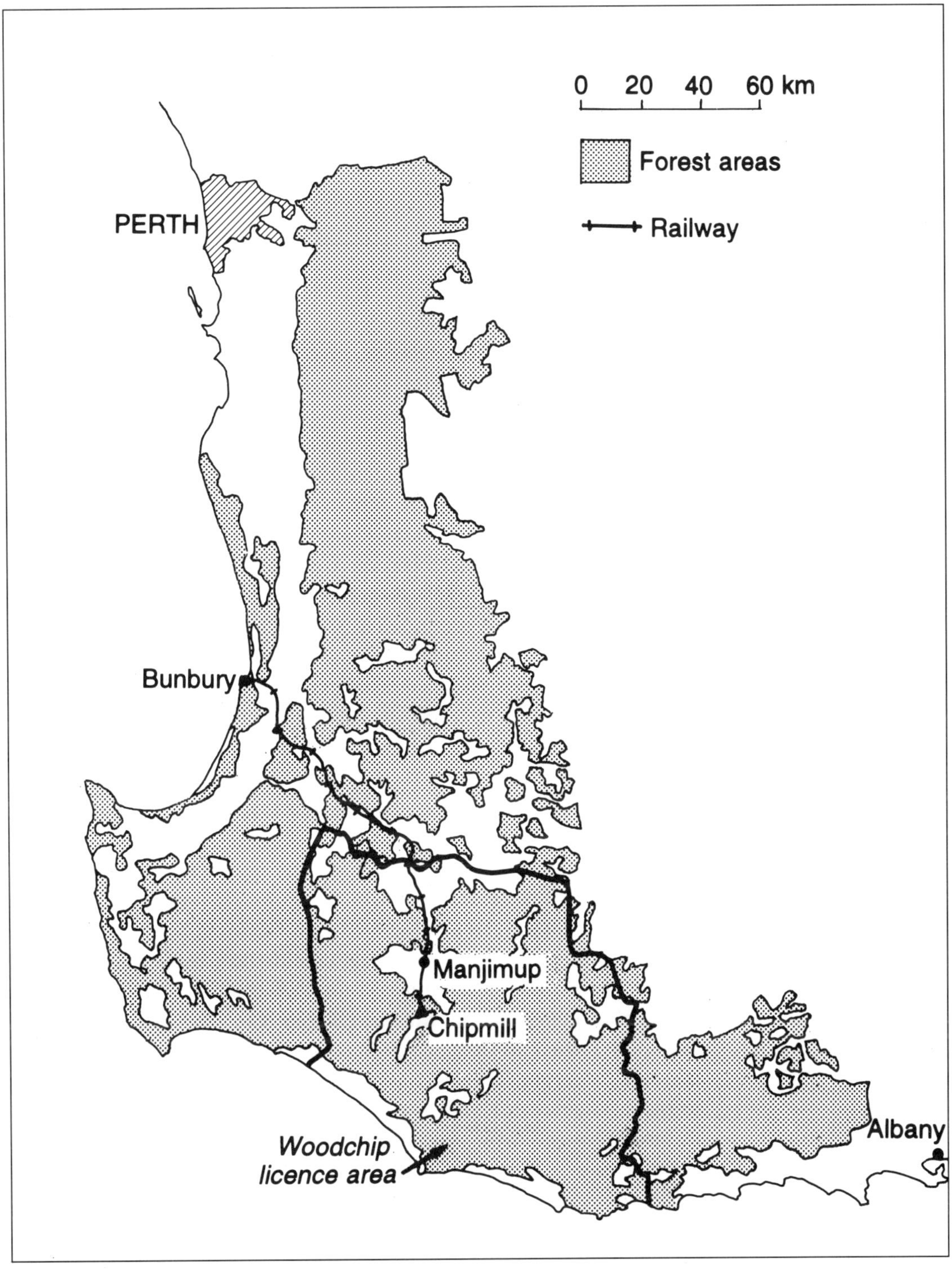

Figure 4.6 South-west of Western Australia, showing woodchip concession

and heads of trees already felled for sawmilling, rather than clear-felling on a broad scale as in the earlier projects. Allen Taylor and Co. Ltd was the largest, and one of the oldest, timber companies in New South Wales; with its string of thirteen sawmills through the State it had best command over such resources. It formed a subsidiary, Sawmillers Exports Pty Ltd, and secured approval to export 350 000 tonnes a year from Newcastle until 1996. The chips were sold to two sogo shosha involved in the initial proposals: Toyomenka which supplied Japan's largest paper company, Oji Paper, and C. Itoh which supplied Chuetsu Pulp Co. The approved quantity was later increased to 500 000 tonnes a year. Like the other companies which started woodchip exports, Allen Taylor was subject to a series of takeovers, ending up as part of Boral.

The second of the later projects was proposed in 1981 and encountered much stronger environmental opposition. It arose when a domestic market for hardwood pulpwood from the Otway Ranges in western Victoria ended. Small quantities had been cut there since the 1950s, first for APM's Maryvale pulp mill and increasingly from 1966 for supplying Smorgon Consolidated Industries' pulp mill in Footscray. When the pulp mill was closed on environmental grounds, Smorgon proposed exporting the Otways wood, with some softwood from pine plantations, from Geelong.[36] The Victorian government promptly called for tenders for 70 000 tonnes a year of logging residues from the Otways, which were won by a consortium of all the hardwood sawmillers in the Otways, Midway Forest Products. The environmental movement raised strong objections; the government changed, and set up an inter-departmental task force to study the problem. Although this recommended using the logging residues for export, the government restricted the sawmillers to exporting only whatever chips they could make from the waste at their mills; the butts and tops of trees they felled for sawlogs had to be left to rot in the forest. However, Midway went ahead and built a small loading facility in Geelong, bought chips from other sawmills, obtained a Commonwealth licence to export 80 000 tonnes a year and entered into a contract with Mitsui for supply to two pulp mills belonging to the Jujo Paper Co. The first chips were exported at the end of 1986. Midway then searched for other wood to increase its output. The patches of privately-owned forest in the Otways were one source for which a licence was obtained for 'silvicultural thinning'. Waste from sawmills located north of the Divide were another source. While Smorgon had a pulp mill, APM had tussled with it for their chips; but when Smorgon shut, APM withdrew and let Midway export them. Wood from land being cleared to plant pines was another source. As these sources were found, Midway was able to have its export licence raised in steps to more than 200 000 tonnes by 1991. To support the expansion, Midway had to install a small chip mill at its

Geelong plant. The wharf facilities limit the size of ship that can be loaded, but they match similar port facilities at Jujo's Yatsushiro and Ishinomaki mills.

The export of sawmill chips fom Queensland was uncontroversial and the Commonwealth approved the export of 300 000 – later reduced to 180 000 – tonnes a year from Brisbane. Australia's expanding pine plantations also provided an uncontroversial resource for export. The possibilities of exports from plantations in south-east Queensland were investigated, and a small quantity of material salvaged from burnt plantations in New South Wales was exported on a short-term basis. The first major project to export up to 400 000 tonnes a year of pine from plantations in the 'Green Triangle' of the south-east of South Australia and the south-west of Victoria was developed by a South Australian company, SEAS–Sapfor Ltd.

Prices and costs

The shipping out of Australia's natural resources did not go unquestioned economically, as swingeing attacks were mounted on tax subsidies and foreign control in the mining industry. For the woodchip industry, the central economic questions were whether the exports were underpriced and whether the royalties collected covered the costs.

Prices

The prices in the rolling five-year contracts for woodchip sales are negotiated between the Australian exporters and the Japanese pulp and paper companies twice a year in the presence of the trading houses – the sogo shosha. APPM negotiates its price first, which then becomes the benchmark for negotiations by the other companies, with adjustments for quality. The various Japanese negotiators are believed to discuss prices beforehand and share information, but this does not alter the trading pattern (Table 4.1). As noted earlier, the Commonwealth checks the prices before renewing the export permits, but makes only a limited comparison with the trends in other export prices, particularly those from the United States. Generally Australian prices have improved relative to those from the United States, and only once has the Commonwealth called for prices to be renegotiated before issuing a permit.[37] In spite of the Commonwealth's checks, questions continued. Several economic studies were conducted which came to broad agreement that Australia was not obtaining prices as high as perhaps it might.[38] There were four issues in this: trends in the real

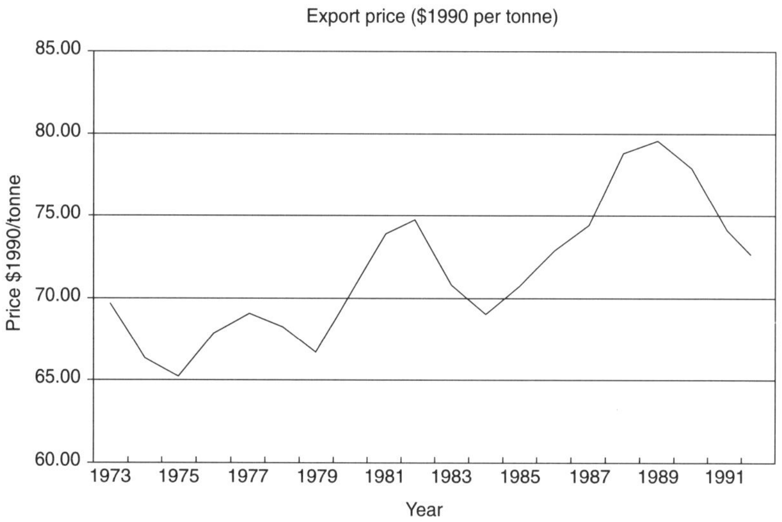

Figure 4.7 Woodchip export prices, 1973–91
Note: Prices shown as three-year moving averages.
Source: Australian Forest Resources.

prices, changes in the exchange rate, competition over the share of the market, and Japanese buying practices and strategies.

The real (i.e. net of inflation) prices received in Australia fluctuated somewhat cyclically, but show no evidence in Figure 4.7 of long-term decline, especially if the first two or three years when the industry was smaller are discounted. Certainly, woodchips fared better than most of Australia's other primary products. While prices did not get worse for Australia, they – like those for other primary commodities – got markedly better for Japan as its exchange rate improved. Australia was able to increase its share of the expanding Japanese woodchip market to almost 30 per cent by 1985, since when it has fallen to a little over 20 per cent as exports of red alder from the United States and beech from Chile have captured most of the expansion. Malaysia, Papua New Guinea, Canada and South Africa are also suppliers, though in a much smaller way.

Although Australian woodchips cost a little more in Japan than those from other suppliers, Japan's general trade strategy of diversifying the sources of raw materials, and the one-to-one link between Australian woodchip and Japanese pulp mills, can be expected both to ensure the continuation of the trade and to prevent any significant increase in Australia's market share. Most studies consider that there should be scope to increase prices somewhat, but given the small contribution of

Australian supplies in the total Japanese fibre furnish, and the structure of the trade, this is unlikely to be achieved.

Costs

Critics also questioned whether the revenues to the forest services were sufficient to cover their costs. Several studies were conducted but were not conclusive: the forest services had little experience in economic analysis, their accounting systems were unable initially to provide economic data, the costs of using public infrastructure (roads, rail, ports) were unclear, the forest operations and costs had not stabilised, and, more importantly, it was not clear which costs the royalties from woodchips should cover.[39] Obviously, royalties should cover the direct costs of the logging operations and some proportion of the new roads built, though these have lasting benefits to others too; similarly, they should cover some of the costs of regenerating and tending the new stands; and they should cover some of the overhead costs of the services; beyond merely covering costs, they might reasonably be expected to show some real profit. Repeated questioning, and the general fiscal crisis of the States from the mid-1970s, led to raising of the pulpwood royalty rates being raised and improvements in accounting practices.[40]

Reorganising the integrated concession regime

Forest resources were supplied to four of the five big woodchip export projects mainly by rapidly extending the integrated concession regime to far greater areas of public forest. As the regime had been successful in laying rights to pulpwood over those to sawlogs during the 1950s and 1960s, it seemed that the woodchip projects would proceed smoothly on their more glorious scale. However, their rapid expansion and far larger scale brought new problems. For the first few years, each project was under great pressure to have enough wood chipped ready for the next ship, whether the roads, the sawmillers or the forest services were ready or not. But even when the early difficulties passed, the balance of interests between the sawmillers and the woodchip companies shifted to what has been described as 'pulpwood-driven' management.

This was as much the legacy of the long history of the sawmills regime as it was the urgency of the hefty demand for woodchips. For example, the forests of the Wesley Vale concession had been cut selectively for well over a century, taking only the best trees at a time. During the boom of the 1950s and 1960s, the sawmills had expanded their production to about twice the level that the forest could sustain in the long term so that, as many of the stands were being cut for the third or fourth time, the average yield and quality of logs were rapidly declining.[41] Just before APPM started cutting pulplogs for its woodchip mill, numerous small logging contractors and sub-contractors were supplying hardwood sawlogs to fifty-six, mostly small, sawmills who held licences (called Exclusive Forest Permits) over sixty-seven different forest areas. The gravity of the resource depletion became inescapable. In 1977 it was realised that although log quotas were reduced by 20 per cent, and further cuts of 30 per cent were scheduled for the 1980s, even more severe reductions would be needed unless all the sawlog operations became fully integrated with the pulplog operations.

Both the Forestry Commission and APPM stressed the complementary advantages of higher productivity and better roads in integrated logging. At first, many sawmillers preferred to continue to log their own permit areas, but such selective logging produced little pulpwood. Irrespective of the sawmillers' preferences, the resource had about eight times more pulpwood than sawlogs, and APPM had to have about four times the quantity of all the sawmills put together; only widespread clear-felling, large operations and substantial roading could obtain it. Clear-felling ensured that every last sawlog was felled, and hence it averted even more drastic reductions in sawmill quotas. The transition to integrated logging was finally pushed through by the Forestry Commission. It replaced Exclusive Forest Permits as they expired with sales contracts which guaranteed sawmillers quantities of logs, but did not provide them with exclusive areas. The sawmillers not only lost their exclusive area rights in the forests, but they also became operationally dependent for their sawlog supplies on logging contractors whose major business was with the woodchip mill. Complaints that the contractors slipped sawlogs into the chipmill were rarely found to have much substance when the Forestry Commission checked them.

The Forestry Commission and APPM planned the operations in a whole series of detailed documents and meetings, coordinated at the top by a Planning Committee of two officers from each organisation, and supported by the work of specialist staff. Although the sawmillers were eventually invited to send a representative to the Planning Committee, and individual millers were invited to collective meetings, they did not have specialist support staff, did not contribute reports to the reviews, and generally relied on the Forestry Commission to protect their interests. It was the demands of the Japanese pulp mills which set the pace; the Australian sawmillers had few options but to follow.

Forest management over much of the four big woodchip concessions was largely 'pulpwood-driven' – in the sense that pulplogs commonly amounted to 80–90 per cent of the total quantity cut in clear-felling. There were several other, less extensive operations. Selective 'sawlog-driven' logging continued in areas such as the Otways, northern New South Wales and East Gippsland, in which pulpwood was produced only from the butts and tops of trees felled for sawlogs. However, the ability to sell such waste and make woodchips from waste at the sawmills doubtless led to more trees being cut. Various other operations, thinning young stands and salvaging wood from areas being cleared for various reasons, added to the total pulpwood available.

Refashioning the forests

The extensive clear-felling on the woodchip concessions radically refashioned the structure of their stands and the landscape of their forests. Although clear-felling for the domestic pulp mills did so too, its impact was surpassed by the size and concentration of the export operations.

Stand structure

Clear-felling is the most drastic treatment. Most, but not all, of the stands in the woodchip concessions had been logged for the best sawlogs, often several times. The roots and crowns of the large old trees which had been left behind had conquered the growing space vacated by the trees felled or fallen; little was left for new young trees, although some came up in the larger gaps, and there were a few maturing trees which had been too small earlier. On the whole, the stands were growing at far less than their potential, causing Australian foresters disparage them as 'depauperate'. Clear-felling, by removing all the old, previously 'useless' trees, provided an opportunity to establish a uniform crop of young, vigorous trees. But it was soon to be found that these 'useless' trees provided abundant habitat for possums and gliders, cockatoos and owls.

Forest landscape

The clear-felled landscape was spectacular and hideous. Its pleasant, green forest had gone; snig tracks, log landings, roads, bare earth, broken trees and charred waste were left. It was a scene of devastation that threatened to extend year by year, or so it seemed, and could be displayed on protesting posters. But the landscape elements altered with time and practice to give ever-changing scenes different from those so threateningly depicted.

The solid roads needed for such heavy, all-weather logging scored the landscape more deeply than before. They were startlingly obvious and their construction was a major source of erosion and stream pollution. However, the size and layout of the felling areas, or 'coupes', had more startling impact. They were large at first; up to 200 hectares in New South Wales, 500 in Tasmania, 200 in Victoria and 800 in Western Australia.[42] Although most were smaller, they were often adjacent to the previous year's cutting so that they joined up to form effective sizes of over 5000 hectares. Most biologists felt that small coupes of 40 or so hectares would be better, but the forest services thought that coupes of 200 or more hectares were necessary in order to have a hot enough fire after logging, at a reasonable cost, to prepare the ground for the next crop. Nevertheless, by the end of the 1970s, more roads had been built and the pattern had changed in response to the environmental objections. The coupes were smaller and widely dispersed in a 'chequer-board' pattern across the forest. For example, at Eden they were reduced to about 15 hectares in 1976, though brought to a more realistic maximum of 60 hectares later. Also, many patches of the forest were left uncut for a host of reasons: to protect the edges of streams and filter any soil run-off from the coupe, to protect soil on steep slopes against erosion, and to provide extra habitat for particular animals, or corridors along which they could move between favoured parts of the forest.

Logging was done more carefully too. No longer were the heads of felled trees allowed in streams, nor could logs be pulled through them. Compacted and bark-strewn log landings were ripped up so that new trees could grow there. Old snig tracks were 'barred' so that water would not scour them. If the soil or roads were too wet, no logging could be done. Against grumble and complaint, such improvements were gradually made and eventually formalised in codes of practice. The landscape was healed most by time; roadsides became less raw, bare coupes were clad in young trees, nature seemed as though it would be as forgiving of woodchips as earlier of goldmining. Yet many environmental changes – particularly to wildlife – were more profound than readily apparent, less healed than they seemed. They remain contentious issues for the future.

CHAPTER FIVE

Forests for Work and Regional Development

The work and wealth that could be produced in rural regions by forest developments, such as sawing timber, establishing plantations or exporting woodchips, were some of forestry's traditional attractions. Colonial politicians had known them well, Hutchins had trumpeted them, and the Development and Migration Commission had realised them in the 1920s.[1] But in the 1970s the simple belief that productive forests attract industries which create employment and other social benefits – a sort of 'sylvan determinism' – had to be reappraised. What changed was that after an unprecedented period of almost full employment since the 1940s, unemployment suddenly soared to levels not seen since the 1930s and, with fluctuations, continued to rise into the 1990s.[2] By 1993 almost 1 million people were unemployed, and perhaps as many again were under-employed. In effect, 10–15 per cent of the labour force was expelled, a pool of people who were only casually employed was created, and many others were discouraged from even looking for work. It was an enduring social problem, the single most prevalent cause of poverty, and a continuing political issue.

The employment crisis commenced as part of the world-wide economic problems of the mid-1970s. Manufacturing employment fell hardest as factories shed labour or closed, in what was a general restructuring of industries. There were important long-term, relative changes too, as both primary and secondary sectors declined relative to the tertiary sector. The remorseless assault of unemployment on domestic and community life was exacerbated by the doctrine of 'economic rationalism' which took hold in political parties and powerfully in the Commonwealth bureaucracy.[3] It held that the country's economic woes would be cured by giving market free rein, cutting government expenditure, and corporatising or privatising many state activities.

The forest sector, so dependent on government, was severely affected by specific resource factors on top of the general trends. During the 1970s and 1980s it shed the jobs of 20 000 people – 37 per cent of its workforce – in spite of starting the whole woodchip export industry, expanding plantations, and increasing production (Table 5.1). However, the full extent and real nature of unemployment was masked from the public eye by the media, which depicted job losses almost exclusively in terms of 'the environment versus jobs'. This deceit was fostered by industry lobbyists and forest services which made, and commonly exaggerated, threats of the jobs that would be lost if more national parks or heritage areas were declared, fewer pines planted, or pulp mill pollution controlled. They twisted the fear of unemployment to advance their own ends in the political mêlée. They conveniently ignored the far greater effects of the incessant, slow grind of capital on labour; 'structural change versus labour' never gained media billing.

The structural changes to hardwood sawmilling and the forest contests mostly occurred in regions whose economic and social base was already declining as a result of great changes in agriculture. They added to what was becoming a grim situation for many rural communities, and sometimes they provided an arena in which anger and mounting frustration could find their expression.

The structural changes to the forest industries were accompanied by substantial changes in their regional location. The woodchip mills were obvious additions in their regions, but for sawmilling there was a gradual but progressive shift of production from the generally over-cut hardwood forest regions to those with increasing plantation resources.

Table 5.1 Workforces in the forest and wood products industries, 1971–72, 1986–87 and 1991–92

Industry	*ASIC Code*	*Number of people employed*		
		1971 1972	*1986 1987*	*1991 1992*
Forestry and logging				
Logging	0303	5 248	5 362	4 127
Forestry	0304	7 080	7 603	5 577
Wood and wood products				
Log sawmilling	2531	15 725	10 535	9 795
Resawn and dressed timber	2532	8 299	6 692	3 848
Veneer and boards	2533	7 498	5 187	4 449
Woodchips	2537	n.a.	790	613
Pulp, paper and paperboard				
Pulp, paper and paperboard	2631	11 297	7 537	6 528
Total		55 147	43 706	34 937

Source: Forestry and logging data are for 1971, 1986 and 1991 from ABS, *Census of Population and Housing.* The manufacturing data are for 1972, 1987 and 1992 from ABS, *Manufacturing establishments* (cat. no. 8203.0).

While the native forests, and hence the hardwood sawmills using them, were naturally strewn around Australia, the plantations were the result of mainly government decisions. Where governments spent their money had been the very stuff of colonial politicking and long argument between city and country. In the 1950s the argument became more formalised under the label of 'decentralisation' policies aimed variously at rural preservation or urban containment.[4] In most States, political lobbying from local members, and the idea that plantations would result in a decentralised softwood timber industry, led to their being widely dispersed.[5] Many of the dispersed plantations proved to be too small to support modern, automated sawmills, and there were considerable advantages of integrating sawmills with pulp or particle-board mills and other factories which could use small trees cut as thinnings from the pine plantations and waste from the sawmills. In the 1970s the expanded planting program was concentrated into a few regions where these advantages could be secured. Thus the transition of sawmilling from native forests to pine plantations involved not only regional relocation, but a shift from a highly dispersed single industry to an integrated one concentrated in a few major regions.

This chapter examines the effects of structural and other changes on employment in the forest sector during the 1970s and 1980s, and its place in rural Australia. The size and characteristics of the sector's workforces are first detailed and the effects of structural changes on employment are described. The chapter then considers the geographical distribution of the sector and its relative importance in local areas. Finally, the values held by people in these 'working communities' are described.

Workforce size

Even the very definition of the forest sector, and hence perceptions of the size of its workforce, was manipulated for political ends in the environmental contests. This could be done because the scope of the 'forest and forest products industries' depends on which of the forest's many products, and how much downstream processing, are included. Conceivably, some apiarists, water board employees or recreation tour leaders, for example, could be included with the loggers and forest workers; and downstream processing could extend from sawn timber to furniture, or from veneer to making matches, or from pulp to printing. But for analysing the consequences of the forest contests, only those activities which depend primarily on the forests should be identified. The dependence varies greatly between those directly affected, such as loggers, and those indirectly or little affected, such as furniture makers. The scope of the sector directly dependent on the forest use and management decisions was defined by the Resource Assessment Commission (RAC) in its inquiry into the sector to include forestry and logging, and the first stage of processing the wood in Australia by sawmilling, woodchipping, and manufacturing wood-based panels, pulp and paper (Table 5.1).[6] This is the best definition that is possible using the industry categories for which statistics are published by the Australian Bureau of Statistics (ABS).[7] Although the 'jobs' argument has been freely deployed in the forest contests, there is surprisingly little information about employment in the sector beyond these statistics.[8]

A much larger scope is frequently cited which doubles the size of the workforce and makes the sector seem more important politically. It includes the secondary processing of wood into joinery, wooden containers, paper bags and cardboard boxes, and sometimes even furniture, mattresses, and printing. The Forestry and Forest Products Industry Council (FAFPIC), for example, used it to claim that the industry employed 106 700 people in 1987, when the RAC definition (Table 5.1) indicated less than one-half that.[9] While the FAFPIC definition properly delineates the council's administrative scope, it includes workers in secondary processing who are far less, or not at all, affected by the contests over the use and management of Australian forests. However, industry lobbyists frequently ignore this; they select the larger scope, and use it deceitfully to inflate the likely extent of job losses from forest use changes or to exaggerate job creations from industrial expansions.

The forest sector is not a large one in the Australian economy. In the RAC definition used here, its 35 000 workers made up only 0.6 per cent of the total workforce in 1991. Other comparisons can be drawn (from the 1986 census): the sector employs about half the number in mining; forestry and logging employ 3.5 per cent of those in primary production; and processing wood from the forest employs 2.8 per cent of the manufacturing workforce.

Workforce characteristics

There are such marked differences in the labour relationships, characteristics and organisation of people working in the sector, that it can be considered to have three distinct workforces: the loggers, the forest workers, and the mill workers. Each contains its own range of occupations, skills, professions, values and beliefs.

Logging[10]

The 4000 or so loggers form the most distinctive workforce. Unlike the others, logging is done almost entirely by some 700 to 800 mostly small contracting and sub-contracting businesses. They are engaged by particular mills to cut logs from the forests and plantations for which the mill-owners have resource rights, and deliver them when required. Under the integrated regime, the contractors deliver the pulplogs to the woodchip or pulp mill holding the major concession, and the sawlogs to several sawmills. The contractors and sub-contractors – mainly log truck operators – are paid negotiated rates on the quantity delivered, and virtually all of the fallers and many of the other workers are paid piece-wages.

In the current terminology of industrial restructuring, logging is a historically 'fragmented' part of the forest products industry which has been generally devolved to small businesses.[11] There are powerful but paradoxical reasons for this and equally paradoxical perceptions of its worth. On one side, mill-owners would find it difficult and expensive to manage operations and maintain equipment scattered through distant forests, or supervise fallers far from the foreman's eye. On the other side, the logging contractors enjoy the feeling of independence in running their own businesses and can profit if they do it well; and the fallers can work their own day and make better money than their cousins in the mills if they are skilful and work hard enough. This attracts them to it, although for many it is the lack of other opportunities and an attachment to place that keeps them to it. But their debts for equipment act as fierce foremen, and although they may make good money while they are young and strong, or when the market is buoyant, they carry greater risks than other parts of the sector.[12] They have some of the highest rates of death, serious injuries and accidents;[13] they are the first affected in the forest contests; the contractors and sub-contractors typically carry large debts for their trucks and tractors, frequently having mortgaged their homes to obtain them; and the sawmills in particular provide little guarantee of steady work. Effectively, they are caught on three sides. In the forest they have no rights to the resource, for the state allocates these to the mill-owners; the mills provide little commercial or job security; and now they face blockages and stoppages during environmental demonstrations. Yet, many have a sense of independence and an attachment to the job that frequently sees sons follow fathers and uncles into the bush; loggers say it is 'in the blood'.

The logging contractors own the bulldozers and loaders, either own or sub-contract the trucks, and employ a few fallers and other workers.

Although there are a handful of large contractors employing 50 to 100 people, more than 80 per cent of firms – those harvesting less than about 40 000 cubic metres a year – employ fewer than 10 people, and usually one of them is the contractor himself who drives one of the machines (Table 5.2). Only 5 per cent of the loggers are classified as managers, administrators or professionals, with a further 5 per cent as trades people. These are some of the lowest proportions in any industry, and presumably such people are employed only in the few large firms. Conversely, logging has the highest proportion of people classified as labourers – 47 per cent. However, such statistician's labels conceal the real skill of the fallers and of those who must be their own mechanics.

Loggers typically work long hours in the bush, far longer than those worked in any other part of the forest sector. According to the 1991 census, logging has only 11 per cent of women workers, of whom about 40 per cent work in the offices of the largest firms. However, the census statistics do not tell the whole story, for in the smaller family businesses, the unclassified work of women is essential to their administration, pay, book-keeping and supplies, while their husbands log the distant forests.

Loggers are the least organised of the three workforces. The timber workers' unions that cover the sawmills and timber yards cover the loggers, but it is hard for their organisers to recruit the scattered crews. Moreover, the inherent nature of piece-work, and the anti-union attitudes prevalent in rural societies and small businesses, often makes them resist joining. Organisers succeed best in the plantation areas and some of the big operations in the pulp and woodchip concessions, where the workers are more concentrated than in the hardwood sawlog operations. Many of the log truck drivers own their own trucks and operate as sub-contractors, while some own more than one truck and

Table 5.2 Proportion of logging firms by employment size

Number employed	*Proportion of firms (%)*
Under 5	62.3
5–9	26.0
10–19	7.9
20–49	2.6
50–99	0.9
100 +	0.2
Total	100.0

Source: ABS, Business Register Counts, August 1992 (pers. comm.)

employ occasional or full-time drivers for them. Hence they are often in ambiguous situations, being sub-contractors, workers and sometimes petty employers. The Transport Workers Union covers the employee and owner-drivers and has been involved in rate-setting with the bigger companies.

The logging contractors, like many of the truck owner-drivers, are almost archetypically independent, competitive, individualistic, and resistant to collective organisation. This is partly because they are widely dispersed, work long hours in the forests, and run their businesses personally. In spite of this, they formed associations at first to try and improve contract rates with the bigger companies, and more recently to counter environmental and government pressures. Some of the earliest were the Maryvale Pulpwood Loggers Association, the Tasmanian Timber and Log Hauliers Association, and the South-East Timber Association which represented contractors to the woodchip mill at Eden.[14] But State and national, rather than regional, organisations were needed to counter the voices of well-established environmental groups. The loggers were tardy in creating them and it was not until 1978 that the Hauliers Association in Tasmania widened its membership and became incorporated as the Tasmanian Logging Association. Under the leadership of Andy Pagett, it nurtured the formation of the Australian Logging Council in 1983. In 1987 a Logging Division was formed within the Forest Industries Federation in Western Australia, and in 1990 the North Coast Logging Association in New South Wales dissolved in order to create the New South Wales Logging Association to which the South-East Timber Association was affiliated.

Forestry[15]

The most distinctive feature of the 5500 forestry workers is that they are almost all employed in large organisations – the State forestry departments and the companies large enough to have their own forests or plantations. There is little evidence that the contemporary trend to 'fragment' the workforce by devolving work to contractors has had much effect, as only 4 per cent of the workforce is classified as employers or self-employed. Employment in the large organisations presumably accounts for the fact that three-quarters of the forestry workforce works only the standard 35–40 hour working week, compared to less than half in logging or the workforce as whole.

Forestry and logging have some similarities, however. Anecdotal evidence that loggers often sought forestry jobs as they got older does not seem to have much statistical basis now; there is little difference in the age distribution of the two workforces, or between them and the

Australian workforce as a whole. Like most rural industries, forestry and logging have a markedly Australian-born workforce; migrants make up only 13 and 8 per cent respectively of workers, compared to 26 per cent in the overall workforce. Also like logging, forestry employs a high proportion of labourers – 38 per cent. The forestry workers are highly unionised, most being represented either by the Australian Workers Union or the Construction Forestry Mining Employees Union.

Forestry is quite unlike logging in that it employs the cadre of professional foresters and managers who, with the technically trained 'para-professional' and supervisory staff, make up slightly over one-quarter of the workforce. This proportion is much like that in the workforce as a whole, but higher than that in manufacturing, and far higher than the 5 per cent in logging. The foresters are of particular interest because they carried the paradigms of forestry into the environmental contests described later. There are probably about 1500 foresters in Australia.[16] They are a highly educated group. All have degrees or 3-year diplomas, and about a quarter now have higher degrees, often taken overseas, which has helped to slightly widen the perspectives of the cadre.

Almost three-quarters of Australia's foresters work in the public sector – a little over half in State forestry agencies, and the rest in other State and Commonwealth agencies. Private industries, mostly the pulp and paper companies, employ 16 per cent, and the remaining 12 per cent are mostly employed in teaching or consulting work. Surprisingly, only about a quarter of Australia's foresters actually manage the forests at the district level, while a quarter are engaged in planning, and another quarter in research and teaching. The emphasis on planning and administration, combined with the traditionally centralised, hierarchical structure of the State forest services, requires almost half of Australia's foresters to live in the capital cities, and a quarter to live in the provincial cities. One paradoxical effect of the environmental contests has been to overwhelm even the district foresters with the burgeoning documentation of environmental assessment, submissions, reports, and community meetings. The time they can spend in the forests has been reduced so much that many can hardly go there any more; they quip that forest officers have been turned into 'office foresters'.

Most foresters belong to their professional association, the Institute of Foresters of Australia, and those in the public services belong to their appropriate State or federal unions. The Institute of Foresters holds conferences, publishes a journal, and makes submissions to numerous government inquiries on the course it thinks that Australian forest policy should take. The cadre of foresters with a sense of a distinct professional identity weakened in the late 1980s and early 1990s as the State forest services were disbanded, merged into other departments or corporatised.[17] Staff reductions and reorganisations severely reduced the number employed, and membership of the professional organisation fell by 8 per cent in as many years.[18]

Mill workers

The workforce in the wood processing industries is much like those in other largely non-metropolitan processing industries, and does not differ greatly from manufacturing in general. Of more importance here are differences between the four major wood industries. They can be seen readily (in Figure 5.1) by comparing the average number of people employed in each mill and the average wage they receive: the bigger, more sophisticated and capital-intense the mills, the higher the wages. These differences echo those in the duration and extent of the resource rights obtained under the various regimes.

The pulp and paper industry's workforce enjoys high average wages due partly to the higher proportions of professional, technical and trades people employed, but also to the industrial relations legacy from the 1950s and 1960s. The unions were able to organise successfully in such large mills, while the companies, well protected in their comfortable monopolies, paid up rather than have their expensive machines idle or lose buoyant sales through strikes. The higher wages won in the pulp and paper mills spilled over into the woodchip mills.

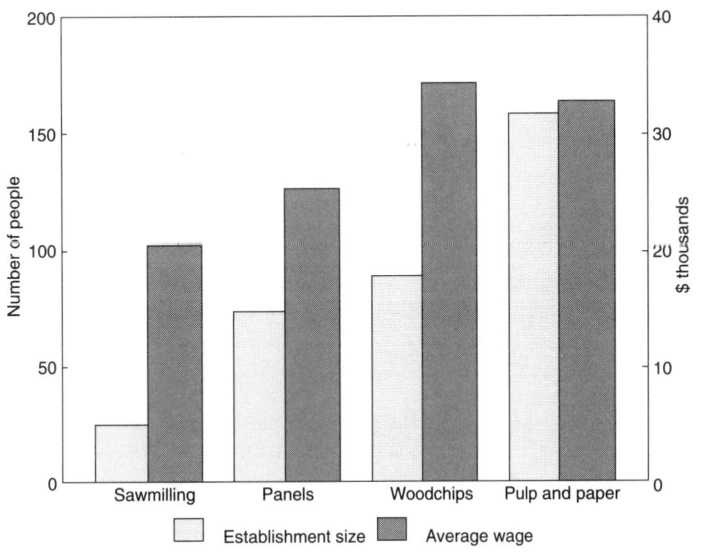

Figure 5.1 Average establishment size and wage in four wood industries
Source: ABS, *Manufacturing Establishments: details of operations (cat. no. 8203.0); Manufacturing Industry: Selected items of data by employment size.* 1987–88 (cat. no. 8204.0).

The sawmilling industry is the least favoured, but even within it, markedly higher wages are paid in the larger mills than in the smaller. There is far less female employment in sawmilling (8 per cent) than in manufacturing generally (27 per cent), and within sawmilling there are slightly more women (10 per cent) employed in the small mills with their lower wages, than in the large mills.

The workers in sawmills and timber yards were covered by the Australian Timber Workers Union, and a local variant in Western Australia. Given the small and scattered mills that made up the hardwood sawmilling industry, and their fluctuating fortunes, it was not a strong or radical union. In the rounds of union amalgamations of the 1990s, it was absorbed into the Construction Forestry Mining Employees Union. The pulp and paper workers were in the Pulp and Paper Workers Federated Union of Australia, which had evolved from an 'in-house' union for APM's workers and was known in the trade as the 'puffer-wuffers', and the more radical Printing and Kindred Industries Union which covered APPM's workers.

Structural changes

The over-arching effect of structural changes during the 1970s and 1980s was that, in spite of the new jobs in the woodchip industry and a 37 per cent increase in the quantity of wood cut from the forests and plantations, employment fell by 36 per cent (Table 5.1). The trends varied in different parts of the sector, as did the causes. Although the environmental contests severely affected employment in some regions, they had only minor influences on the national trends.

Mechanisation in logging and forestry

Logging and forestry, like other labour-intensive rural industries, continued the long-term trend of mechanisation to the extent that employment dropped even while production increased. This is easily seen in logging by comparing employment to the volume cut, although it appears that the historic trend may have weakened in the mid-1980s. In forestry it can be seen by comparing employment with an index representing the wide variety of tasks that forestry workers do: regenerating cut-over forests, establishing new plantations, fighting fires, tending young trees, building and maintaining forest roads, and so forth (Figure 5.2).[19] The rising productivity of labour inherent in the trends is only part of the economic picture in which the efficiency of the other inputs – capital, land, and management – should also be

Valmet 911 Harvester machine, driven by Mark Blackberry, thinning a 17-year-old pine plantation at Laurel Hill, NSW.
(Mike Jones, Sisu Logging Australia Pty Ltd)

considered. The data do not enable this to be done, although for plantation forestry the prospects for greatly increased productivity from the land are bright.

The larger scale of logging required to supply the woodchip and new softwood sawmills enabled a number of contractors to expand their

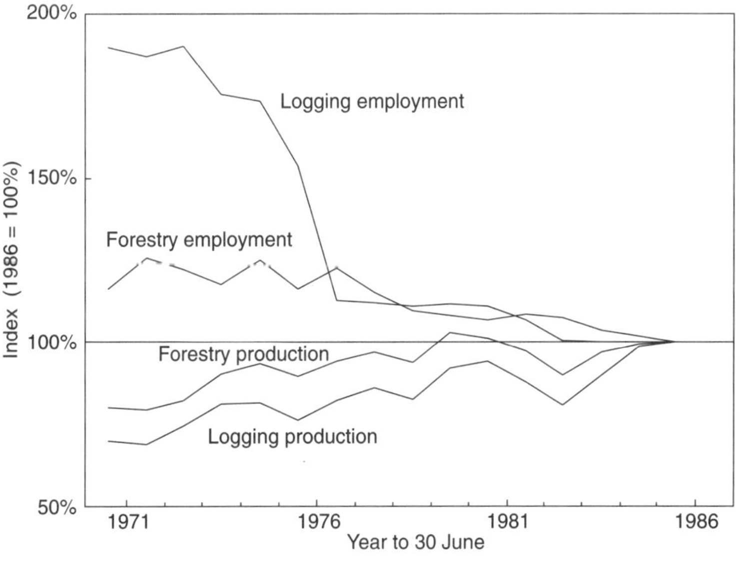

Figure 5.2
Employment and production in logging and forestry, 1970–90
Source: ABARE, *Australian Forest Resources*, various years.

businesses considerably. By 1992 twenty-nine of them were logging more than 100 000 cubic metres a year.[20] Far more specialised machines were introduced too, especially in the plantations where feller-bunchers and ingenious harvesters were even able to fell the trees. The proportion of the workforce operating machines rose rapidly during the 1980s from 13 to 34 per cent, and more qualified trades people had to be hired. More women entered the industry too, most in the offices, but a slowly increasing number in the forest operations; even so logging remained a largely (90 per cent) male industry.

Forestry also became much more mechanised. The era of hand work ended, except where – on jobs like planting trees on steep slopes – it could not be avoided. In both forestry and logging the statistician's classification of many workers as 'labourers', rather than as 'plant and machine operators', disguises the fact that many 'labourers' spend much of their time operating small machines. The proportion of women in forestry rose markedly during the 1970s and 1980s from a meagre 4 to 14 per cent due mostly to more women working directly in the forests. There were more women foresters too, but they still make up only about 8 per cent of that profession.

Concentration and centralisation in sawmilling

The widespread failure of small businesses, and the concentration of ownership – through takeovers and mergers – into larger companies, which then centralise their production into fewer sites, are common features of hard times. They were magnified in hardwood sawmilling by the inexorable reductions in the quantity of sawlogs which could be cut from the native forests. Reductions to achieve a long-term sustainable yield, such as those discussed for Tasmania, were widespread and it became clear that more would be needed. In Victoria the quantity of hardwood sawlogs which could be cut was forecast to fall by 25 per cent by 2010, but in some regions, like East Gippsland, log quotas had to be reduced by 50–60 per cent.[21] In New South Wales, the progressive reductions were expected to have finally been brought down to the State-wide sustainable level by 1998.[22]

The consequences differed. Many small hardwood sawmillers found that their reduced log quotas were too small to be profitable. This, combined with tighter markets and the depressing atmosphere of endless environmental conflicts, induced many to leave the industry. Their exit was eased because the large sawmillers were keen to buy other mills purely for their log quotas. They could then close the purchased mills, transfer the quotas to their main mill, and thus maintain their production and markets. On top of this, some of the

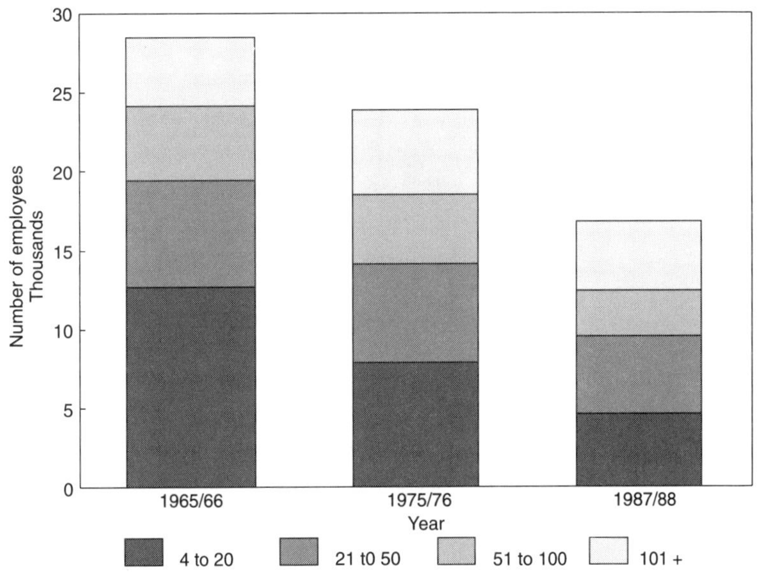

Figure 5.3 Sawmill employment by size class, selected years
Source: ABS, *Manufacturing Establishments* (cat. no. 8204.0).

large sawmilling companies were subjected to mergers and takeovers which left them as part of large multi-sector corporations. This changed hardwood sawmilling from an industry characterised by family businesses to one of companies and corporations. Notably, jobs were shed in all size classes of sawmills, but most severely by the small mills employing less than twenty people.

Plantation transition

While the small hardwood sawmills were closing, the softwood sawmills, less harried by the environmentalists, were thriving on the expanding and fattening pine plantations (see Figure 2.1). Large enough resources had been created in the concentrated plantation regions to supply the largest sawmills; by the 1990s there were sixteen softwood sawmills – but only one hardwood mill – cutting more than 75 000 cubic metres of logs a year.[23] The big softwood sawmills were highly automated. They were equipped with handling systems for sorting their streams of logs into uniform batches which could be sawn and resawn into standard sizes in computer-controlled processes. This contrasted sharply with even the best-equipped hardwood mills, in which every log had to be considered separately if the best timber was to be recovered. There were differences in the markets also. The softwood mills sold timber that was kiln-dried, dressed, strength-tested and graded, in accurately sized packs with standard dimensions, through marketing organisations. They captured much of the market for house framing

Modern pine sawmill. SEAS Sapfor Ltd's mill at Tarpeena, SA, showing Kym Harrup operating an automatic stacker.
(SEAS Sapfor Ltd)

timber which had been supplied with green, rough-sawn timber from the small hardwood mills, often directly to the builders. Some of the hardwood mills improved their processing and concentrated on particular markets in which the inherent strength or appearance of the eucalypts was of advantage, but they were always left with a proportion of low-grade material which was difficult to sell.

In spite of the economies of scale and high degree of automation in the softwood sawmills, the smaller sizes of their logs and the extra processes they adopted to prepare their timber for market meant that they needed more labour than the hardwood mills. For example, the South Australian sawmills, which relied entirely on plantation pine, needed seven people for every 1000 metres of sawn timber that they cut each year, whereas neighbouring Victoria, which relied on its native forests for a little over half of its production, needed only four.[24]

Structural change in the pulp and paper industry

The comfortable oligopoly of the pulp and paper industry, with its tidy divisions of the market among the few large companies, temporarily broke down in the less comfortable economic circumstances of the 1970s and 1980s, only to be restored in the 1990s more tightly than ever. As with its creation, both international and national interests were involved. The largest paper-maker, APM, set the pace, changing its name to Amcor Ltd and embarking on a vigorous program of diversification, vertical integration and expansion. It absorbed Smorgon's paper-making operations and bought several packaging companies using paper as well as other materials. It bought an Australian wholesale paper merchant, Edwards Dunlop, in 1988 and took an 18 per cent share in the British-based multi-national paper distributors, Spicers Ltd, in 1991. These strategic investments helped Amcor, which had built a new mill at Maryvale and upgraded its older paper machines, to capture almost one-third of the fine paper market – particularly the thriving market for copying paper – against competition from NBH's subsidiary, APPM.

APPM, the traditional maker of fine printing and writing paper, had intended to build its 'world scale' pulp mill at Wesley Vale, which might have led it to modernise its paper-making too. However, this had not succeeded and it was left, in the words of one commentator, as 'an industrial dinosaur, its machinery ... ancient and hopelessly inefficient'.[25] Its mainland paper merchants were no longer able to sell all its production against Amcor's better quality. The situation could not go on, and in 1993 Amcor bought APPM's pulp and paper mills at Burnie, Wesley Vale and Shoalhaven, and most usefully its distribution and marketing network for fine paper. Amcor then started to upgrade some of the old paper machines it had bought, close others and improve the environmental standards. In the process, it reduced employment by 530 jobs.[26] The takeover left NBH with its forest concessions and plantations in Tasmania from which it continued to supply both the mills it had sold to Amcor, its two existing woodchip export mills, and a new one it set up at Burnie to export wood from its eucalypt plantations.[27]

ANM continued as the sole newsprint manufacturer, doubling its production, but changing its ownership. Wood resources in Tasmania limited the possibilities of expanding its Boyer mill, so, in 1982, ANM built a new mill at Albury based on wood from pine plantations in New South Wales and Victoria. The new mill, modern, efficient and automated, had to carry higher charges for pollution control, wood transport and power, but used only about one-quarter of the labour of the old Boyer mill.[28] ANM's joint Australian owners – the Herald and Weekly Times, and the Fairfax newspaper group – lost control when the former was taken over by Murdoch's News Ltd, and the latter sold to the New Zealand multi-national corporation, Fletcher Challenge, in 1988. The environmental thrust to increase the amount of paper recycled led to the development of processes for 'de-inking' the old paper. This not only enabled ANM to build a newsprint recycling facility at its Albury mill in 1993, but also encouraged a smaller packaging firm, Pratt Industries, to plan its own facility in Sydney.[29]

The old Boyer mill needed substantial investment and radical improvement if it was to continue. It had almost cut out the resource of old-growth ash trees in its Florentine Valley concession; its wood-handling system, which reduced large logs to small billets before grinding them into pulp, was extremely labour intensive; its chlorine bleaching process used mercury cells which were only able to be operated under Ministerial exemptions from the environmental regulations; and the mill ran at a loss – $18 million in 1989.[30] A new thermo-mechanical pulp mill was installed in 1987, the wood input was changed to receive only thinnings from pine plantations, regrowth ash and chips from sawmill waste, a chlorine-free bleaching process was installed, recycled paper was added to the furnish, two paper machines were modernised, an old one was closed, and new work practices were negotiated with the unions. Overall, the workforce needed to make a tonne of paper was reduced from eight to five people.

Shedding mill labour

Employment was reduced by these structural changes, and some regional effects from environmental disputes. As well, jobs were shed in sawmilling and making pulp and paper by investment in mechanisation and automation, removing demarcations between trades, 'multi-skilling', contracting maintenance out to smaller firms, simplifying jobs, and just making the crews work harder. The combination of effects can be seen by comparing employment with production (Figure 5.4).

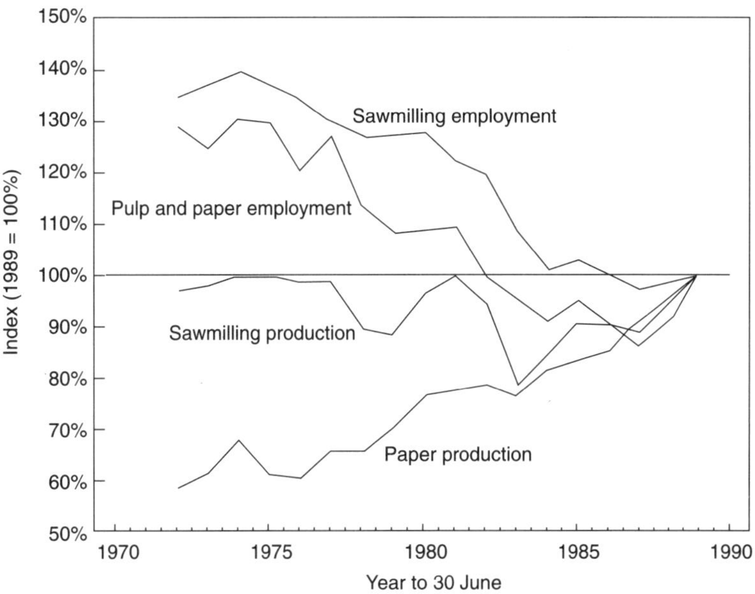

Figure 5.4
Employment and production in sawmilling, and pulp and paper making
Source: Australian Forest Resources.

Although both sawmilling and pulp and paper making shed labour during the 1970s and 1980s, their production differed markedly. Sawn timber production and consumption fluctuated with the peaks and troughs of the building industry, but showed no long-term trend to increase. Nor was domestic production able to replace imports, which continued to supply up to 30 per cent of the demand. By contrast, the consumption of paper and cardboard increased endlessly in spite of economic circumstances. The per capita consumption of 120 kilograms a year at the start of the 1970s rose to 160 by the end of the 1980s – one million tonnes a year more. Imports rose from 500 000 to 930 000 tonnes a year, which added to Australia's severe balance-of-payments problems in the 1980s. But production increased far more, by 825 000 tonnes. The trends varied between commodities. The packaging sector even managed to raise its exports above imports in the 1990s; newsprint imports were able to be slowly reduced; but imports of printing and writing paper increased to 490 000 tonnes a year with negligible exports to offset them.

Even though the total production of paper increased by 75 per cent during the 1970s and 1980s, employment was reduced by 23 per cent; whereas it had needed 10 people to make 1000 tonnes of paper at the beginning of the period, only 4.6 were needed at the end. Old, narrow, slow paper machines were shut as new, wide, fast ones opened.

Computer control of the big paper machines became standard, which also ensured better quality and enabled lighter grades of paper to be used in many applications. Unlike logging, forestry, and sawmilling, the level of female employment in the pulp and paper industry dropped from 10 to 7 per cent. This was mainly because the labour-intensive methods of cutting, counting, and wrapping reams of paper for sale were automated and the work of many, mostly women, in the 'finishing rooms' of the paper mills was ended, a situation exacerbated by male pressure from a right-wing union in at least one site. Overall, the pulp and paper mills were solidly unionised and most of the reductions in employment occurred more or less smoothly, either by not replacing retiring workers or through negotiated redundancy packages. The exception was at NBH's (APPM) old Burnie mill where a bitter, month-long strike took place in 1992. While it is clear that at the national level, restructuring and the continual shedding of labour were the major reasons for the decline in employment in the wood industries during the 1970s and 1980s, their effects, and those of the environmental contests, were felt unequally in the forest regions.

Rural societies and timber towns

The natural distribution of Australian forests, the geography of land settlement, and the history of forest reservation left most of the originally forested districts with a mix of agricultural and forest land. Generally, forests had been reserved only after the best-watered, most fertile and flattest land had been grabbed for agriculture, so that now a typical rural landscape has agriculture in the foreground, forests on the hills behind, and perhaps some plantations in the middle distance. It is dotted with small settlements and towns, occasionally provincial cities, created directly or indirectly for those who worked the land and forests. In this sense, rural Australia is one of 'working communities'.

In most parts of Australia, the wood industry is a relatively minor component in many, primarily agricultural districts, although in a few it is much more important. Some, such as southern Tasmania and the south-west of Western Australia, still have a high a proportion of good-quality native forests; some, such as the south-east of South Australia and the Oberon and Tumut districts of New South Wales, have extensive pine plantations; and the 'paper company towns' such as Burnie and New Norfolk in Tasmania, Traralgon in Victoria, or Millicent in South Australia have substantial employment provided by their large mills.

The published census statistics that are readily available for Australia's local government areas illustrate the distribution of the logging, forestry, and wood-processing workforce – called the 'timber industry' in this discussion – but not of the pulp and paper workforce, which is included in general manufacturing.[31] Limitations of the data, definitions, and variations in the size of local government areas between States prevent this approach yielding more than a very general picture. The importance of the timber industry in each local

Table 5.3 'Timber towns' and 'paper company towns'

Local Government Area	*State*	*Timber workforce 1986 (%)*	*Change 1986 to 1991 (%)*	*Main resource*
Timber towns				
Nannup	WA	32.7	−27	Forest
Manjimup	WA	26.6	−18	Forest
Spring Bay	Tasmania	23.0	−18	Forest
Penola	SA	21.3	−19	Plantation
Oberon	NSW	19.0	+4	Plantation
Bombala	NSW	18.4	−17	Plantation
Mount Gambier	SA	17.5	−21	Plantation
Orbost	Victoria	17.5	−17	Forest
Tumut	NSW	17.0	−13	Plantation
Ringarooma	Tasmania	14.6	−17	Forest
Scottsdale	Tasmania	13.7	−6	Plantation
Myrtleford	Victoria	12.9	+12	Plantation
Omeo	Victoria	12.9	−80	Forest
Tumbarumba	NSW	12.7	+10	Plantation
Esperance	Tasmania	12.4	−30	Forest
Fingal	Tasmania	12.2	−17	Plantation
Beachport	SA	12.0	−8	Plantation
Circular Head	Tasmania	11.4	−31	Forest
Donnybrook–Balinup	WA	10.5	−39	Forest
Paper company towns				
Millicent	SA	8.0	−27	Apcel mill
New Norfolk	Tasmania	5.9	−20	ANM mill
Burnie	Tasmania	3.6	−22	APPM mill
Traralgon	Victoria	2.8	−9	APM mill
Albury	NSW	0.6	−14	ANM mill

Source: Calculated from ABS, *Census of Population and Housing*, 1986, 1991.
Note: 'Timber workforce' includes forestry, logging, wood, wood products and furniture but excludes pulp, paper and paperboard.

government area can be gauged by its proportion in the total workforce. Ranking Australia's 900 local government areas according to this proportion enables the most important to be identified.

The 'timber towns' group of nineteen local government areas, in which the timber industry makes up more than 10 per cent of the workforce, is listed in Table 5.3 with the 'paper company towns', in which the pulp and paper mills require a workforce additional to those reported for the timber industry. Only in the top four 'timber towns' is timber the largest employer, while in the group as a whole, agriculture is the largest single industry. The forestry and logging workforce is about one-third the size of agriculture's, but the wood mills provide about two-thirds of the manufacturing employment. Half the timber towns rely on plantations to supply logs to their mills; the other half rely on the native forests, which makes their economies especially vulnerable in the environmental contests. Only three of the timber towns, all dependent on plantations, increased their timber workforce in the late 1980s; all the rest lost employment.

Beyond the timber towns are an important group of thirty-eight local government areas in which between 5 and 10 per cent of the workforce are in the timber industry. This group contains some major forests about which significant environmental contests have occurred. It includes the north coast and south-east forests of New South Wales, the coastal and tableland forests of Queensland, forests in northern and central Tasmania, and the Otway forests of Victoria. In this group, the forestry and logging workforce is only one-tenth the size of agriculture's, but the wood mills provide about one-third of what manufacturing employment there is.

Generally, Australia's forest regions are poorer than the rest of rural Australia. Both groups of local government areas have higher proportions of unemployed people and households on very low incomes than non-metropolitan areas generally. The Otway forest region was shown to have fewer migrants and a lower participation by women in the workforce than in the rest of non-metropolitan Victoria, and this is probably typical of the hardwood forest regions.[32]

Many small settlements which had grown up or expanded around the small hardwood sawmills are also typical of these regions. There is no overall statistical analysis, but the closure of small hardwood sawmills from the 1970s led to the virtual demise of many dependent settlements. Some consisting only of a handful of unprepossessing mill houses were, perhaps, no great loss; but in others the departure of the mill families, on top of departures of farming families, meant that they could no longer support a school, a shop, or a post office, for example. Thus the structural changes which have been discussed nationally, had their effects not only between regions and local government areas, but also refashioned the human landscape in a detailed way within them.

The distribution of the wood industries had several implications for the social impacts of the forest contests which emerged. Unlike Canada or mining for example, Australia had no large, single-industry towns dependent on the forests. However, this was less of an advantage than it might have been in the hardwood forest regions because agriculture, the major sector, experienced a series of economic crises, droughts and other difficulties in the 1970s and 1980s. Thus the wood industries often provided what little diversification there was in their typically narrow rural economies which had little ability to offset the consequences of the forest contests. Some of the coastal areas were better placed with fishing and tourism, but neither offered jobs to displaced timber workers. Their populations were also growing from the migration of urban people for retirement, work, or an alternative lifestyle in pleasant, warm places such as the north coast of New South Wales.

Unlike most of the hardwood forest areas, a few areas thrived by exporting woodchips or using their plantation resources. Eden, which had been such a sleepy backwater in New South Wales in the 1960s, became a busy centre boosted by its expanding tourist industry as well as its woodchip mill. Mount Gambier, Dartmoor and Millicent in the 'Green Triangle' region in the south-east of South Australia and south-west Victoria did well from their plantations. Not only were sawmills, particle-board and paper mills expanded, but processes were added to make specialised products like laminated beams. The Oberon, Tumut, and south-east Queensland regions are developing similarly, though they are not yet quite so diversified.

Multiplying the impacts

Thriving industrial developments, such as those in the plantation and woodchip regions, provide the classic advantages of economic development: direct employment, indirect employment through the work needed to supply the mills with the whole range of goods and services they need to keep going, and employment that is induced through the spending of shareholders, workers, and suppliers. These effects 'trickle down' through the whole economy. In the expansionary period of 1950s and 1960s economists were able to build models – called 'input-output matrices' – of whole economies, and estimate by how much the effects of investment in a particular sector would be 'multiplied' by its linkages to other sectors.[33] This was particularly attractive to the forest services, which needed to wheedle more money out of the Treasuries for plantation expansion, and justify the low royalties charged to the woodchip and other industries. Their case had never been an easy one to argue: investments in plantations would not yield a return for twenty or thirty years, the royalties for the wood were set by administrative and political processes rather than market forces, resource knowledge was poor, and future demands uncertain. In this situation, the income and employment multiplier factors for forestry could be deployed to justify investments in forestry on the grounds that they would stimulate later economic and regional development. For example, the income and employment multipliers were calculated for a large plantation development project in south-east Queensland, which was said to provide 'an excellent foundation for decentralization and the relief of unemployment'.[34] The same techniques were applied to provide realistic estimates of the employment effects of the woodchip industry after wildly optimistic forecasts had been made by the Minister.

As the environmental contests arose, the forest services and industries found that they needed to defend 'their' state forests against transfers of land to national parks, or restrictions on their use as a timber resource. Again they deployed the multiplier factors to estimate the regional effects of such changes. In an early example in 1976, Kennedy and Lodge calculated the employment effects if the Gembrook forest area in Victoria was used as a park instead of for timber production and a mix of other uses.[35] The size of the multiplier factors was taken to be of such importance in both the advocacy and defence of forestry, that several further studies were conducted and the Standing Committee of the Australian Forestry Council ordered a review of all the factors in use.

The technique suffered from considerable technical difficulties. First, it was based on 'point estimates' of the economy in particular years only, and hence it ignored the whole process of structural change and job-shedding which was having such an effect on the wood industries. Moreover, the particular years for which estimates were available were not always typical. Second, reliable data were available only at the national level. Attempts to disaggregate them to State or regional levels faced considerable expense in collecting local data, and were far less reliable. Third, the technique was designed for the analysis of changes at the margin and was inappropriate for larger or cumulative changes. Although some analysts produced simply ridiculous results, it was capable of providing substantial insights into the operation of regional economies, such as Oberon's, if sufficient local data were collected.[36] Far more serious difficulties came not from the methods of economic analysis, but from the manner in which the information was used to exaggerate the impacts of changes in forest land use, and camouflage the effects of structural changes.

Valuing forest work and rural life

The environmentalists continually asserted that the way in which the forests were being fashioned was wrong, not just because of its consequences, but because the values which motivated it were wrong. Although their moral virtues of 'deep ecology' and an 'ecocentric' philosophy were proclaimed loudly in the forest contests, the values held in rural working communities were rarely understood or respected.

The social nature and values of rural life, long the subject of myth and stereotype, have been probed in several detailed studies of particular communities, although none are specifically of timber towns.[37] Additionally, some information on timber towns has been gathered for Environmental Impact and similar statements. Three strands, relevant to the environmental contests, emerge as likely to apply widely. First, although the fundamental dimensions of class, status, and gender determine the real structures of power, social cohesion is highly prized; differences are not flaunted in dress or leisure, friendliness is valued, personal and family networks are close and of long standing. Second, social cohesion is reinforced by the marginalisation or exclusion of non-conforming individuals. Newcomers from the cities, even in traditional occupations like school teaching, may find it hard to be accepted at first, while radicals in dress or occupation are rejected. Sharply polarised divisions are created in areas, such as the north coast of New South Wales, where large numbers of highly diverse newcomers arrived and some very different settlements, such as that at Nimbin, sprang up. All the newcomers tended to be stereotyped as highly undesirable 'bludgers', 'druggies' and the like.[38] Third, rural people have a strong sense of belonging to their place, not only from long residence, but importantly as contrast to the remote, unfriendly city, the seat of uncaring governments, big business, big unions, in hard times the source of woes, and now it seems of environmentalists. And there are hard economic reasons too; those who lose a job lose cheap housing too if they move to the city, or if they own their house, they fear that they could never afford another in the city.

People working in the timber industry commonly come from rural or timber industry backgrounds. It was common for workers in the hardwood sawmills to move from mill to mill with the ups and downs of the trade, or simply in search of change. A social survey of the sawmilling town of Heyfield in Victoria found that the workers often have a strong sense of identity with 'their industry', analogous to the general rural sense of place.[39] Men identify themselves with 'the bush' which they are loath to leave, but women's lives in such centres are restricted by poor transport, the lack of community facilities and most of all the lack of work; they feel isolated. For young women, and increasingly for young men as the hardwood industry declined, this lack of opportunities was critical.

The values held by workers in the forests and sawmills of the north coast of New South Wales were presented, in an incisive study by Watson, as originating in working-class as well as rural culture.[40] There were many ambiguities, as with any human group, but a central outlook related personal worth to productive work in a practical and material sense. Typically, knowledge and skills in the practical tasks of production – expert tree falling, accurate sawing, engine repairs, for example – were thought of as more useful than theoretical knowledge; while strength, endurance, and the ability to work hard and long were the identifiers of masculinity. A similar outlook pervaded their valuation of the forest based on intimate, local knowledge, sometimes stretching back one or two generations. They had fashioned the forest, it was the evidence of their work. It was valuable for what it had produced and what it could produce. Waste was deplored, when looking either back to past, less careful times, or to present transfers of forest to national parks. The environmentalist claims that forests should be removed from timber production were thus objectionable to the timber workers on several grounds: they were an immediate threat to livelihoods, they were morally wrong as waste, they denied the worth of their past lives in fashioning the forests as they had, they were seen as part of the urban domination of rural life, and they were commonly advocated by people they rejected.

An assembly of values

This chapter has shown that the forest sector contained what can be considered as an assembly of disparate values: some inherently contradictory, some cooperative, others unifying. The advent of the environmental contests masked contradictory values and highlighted unifying ones. The great structural changes in the industries, the demise of small settlements, the unemployment of 20 000 people occurred with rarely a whimper from small firms, unions, or local governments. The difficulties of the overall and agricultural economies could also be blamed, yet the spotlight was kept only on the environmental contests.

The essential interests of capitalists are not those of workers or rural communities, who benefit only from what firms cannot avoid spending on wages, materials, services, royalties, and taxes. Especially in hard times, firms must reduce what 'trickles down' from their businesses wherever they can if they are to survive against their competitors. Moreover, the interests of different groups of capitalists conflict over resources, prices and markets. Yet all must cooperate, for they depend on each other: shareholders on mill workers, mills on logging contractors, contractors on fallers, and in the end all on the forest and its workers. They are thrown together by necessity, but are also bonded by shared values; at best, virtue in production, pride in the products, and conservation as wise use of the forests. All these were assaulted.

CHAPTER SIX

FORESTS FOR EXISTENCE AND EXPERIENCE

During the 1970s large numbers of people started to prize the forests for their very existence as environments and for the spiritual experience they offered as wilderness. These values had ancient roots in Western culture, but it was their widespread endorsement which made them new for public policy; they were asserted in contradiction to the long-established values of development, and conservation as wise use. The new values, which came to be called 'environmentalism', were the essence of the forest contests which ensued, but were far from clear-cut in origins or expression; rather they were a diverse and evolving set of beliefs, claims and practices which surprised and confused the established order.[1] The difference was substantial; the forests were being valued not as resources by the various capitalists and workers involved, or by the consumers of their products, but for intangible values by people who did not benefit in a material, or often even in a personal way, yet who were passionate and articulate about them.

The questioning and the questioners were highly diverse: concerns ranged from local to national, or even global; attitudes from conservation as wise use to complete preservation; politics from conservative to anarchist to socialist; strategies from reform to radical change; actions from letter-writing to peaceful protest to industrial sabotage; and organisation from individuals and small groups to peak national bodies able to pressure governments. Ecology flourished as natural scientists piled on hard questions and found funds for their studies. Interactions with Aboriginal rights and feminism further diversified the scene. Collectively, the diverse questioners, activists, scientists, groups and bodies involved are known as the 'environmental movement'.

Context and conjuncture

The environmental movement developed in Australia as an intimate and vigorous part of Western thought. Some of the gross environmental impacts were obvious causes, but social changes from the 1960s were clearly important as the context which shaped its social and political expression as one of several 'new social movements' operating in contemporary politics.[2]

Context

Outdoor recreation

The most obvious and least controversial change was that more cars, better roads and the network of logging tracks snaking into mountains enabled far more people to take more outdoor recreation and see more of Australia's forests than ever before. During the 1970s the number of people going bushwalking increased at 7 per cent a year and the membership of four-wheel drive clubs increased at 10 per cent, for example.[3] By 1986, more than 4 million people a year, or about one-third of the population over fifteen years of age, visited national parks.[4] Pleasant, rather passive and social activities, such as pleasure driving, picnics or barbecues with friends, and short walks were chosen by most people, but more energetic activities such as bushwalking, camping in the wild, and mountaineering were also more widely undertaken.[5]

The people visiting the national parks were found on average to be better educated and younger, but not to have higher incomes than the general population.[6] Those who engaged in the more energetic activities and belonged to bushwalking and environmental groups were found to be largely from 'a well educated, highly paid, professional elite, with beliefs and attitudes reflecting a comparatively strong sense of their own political efficacy and concern for the preservation of the natural environment in an undisturbed condition'.[7] This association of environmental attitudes with a particular social location was particularly important because of the expansion in what came to be called the 'new middle class'.

New middle class

The ascendancy of employment in the tertiary sector, the increasing sophistication of manufacturing, and rising public employment changed the balance of society.[8] The 'new middle class' became a label for the increasing number of people who were typically urban and employed in the tertiary sector. Their jobs were frequently in or reliant on government, and they were distant from any direct experience of primary or secondary production, well educated, and skilled in white-collar or highly technical work. Such people could not be placed neatly on the historic divide between capitalists and workers. This became politically important as their numbers, particularly in marginal urban electorates, could sway the fortunes of political parties.

Public participation

People wanted to know more about and sought more say in public affairs, forcing reluctant governments to set up more information services and various forms of 'public participation' in inquiries and planning processes. Generally, this involved five steps: publishing a factual document describing a proposal, receiving submissions from the public, publishing a draft plan or decision, receiving more submissions, and eventually issuing the final decision or plan. Although the accuracy and adequacy of all these steps was often contentious, and the influence of the submissions on the decisions problematic, public participation became an established feature of decisions about land use and forest management; no longer could such matters be kept entirely within the bureaucracy.[9]

Back to the land

A movement of people out of the cities and 'back to the land' commenced in the 1960s. Most noticeable was the sprawl of 1 to 5 hectare semi-rural subdivisions around the cities for families who could afford them; further out were larger blocks for weekend retreats and hobby farming; and further still, commercial plantations started by wealthier professional people for rural association or taxation advantage. But it was the move of people seeking an 'alternative lifestyle' which affected the forest contests. Some were individual 'new settlers', or 'blockies', who took on small, often marginally productive blocks of rural land, and endeavoured to live by what they could produce. Others were groups of people seeking communal ways of living, typically on cheap land so degraded that it was unsuitable for agriculture, yet was beautiful, close to forests and in need of care.[10] Both were strongly imbued with the ideals of environmentalism, and their settlement in regions such as the north coast of New South Wales provided a local nucleus for forest contests.

Conjuncture

The origin of the forest contests can be traced, for the purposes of this chapter, to the conjunction of these general changes in Australian society, rising environmental concern, issues in Victoria and Tasmania which became politically important from the end of the 1960s, and the assertion of a national interest in the environment.

Table 6.1 Chronology of some environmental events, 1965–83

Year	*Base*	*Event*
1965	National	Australian Conservation Foundation formed.
1967	Tasmania	Save Lake Pedder National Park Committee formed.
1968	Victoria	Little Desert land clearing opposed.
1970	Victoria	*Environment Protection Act.*
		Land Conservation Act.
1971	Commonwealth	Department of Environment, Aborigines and the Arts created.
1972	Tasmania	The United Tasmania Party formed as the world's first Green party.
		Lake Pedder flooding proceeds.
	Commonwealth	Whitlam Labor government elected.
1973	NSW	Green bans placed by labour unions in 1971 on demolition in the Rocks area of Sydney.
	National	Conservative ACF Council replaced by moderates and activists.
1974	Commonwealth	*Environment Protection (Impact of Proposals) Act.*
	Commonwealth	Committee of Inquiry into the National Estate reports.
1975	Commonwealth	*Australian Heritage Commission Act* set up the Commission and opened the Register of the National Estate.
	Commonwealth	*National Parks and Wildlife Conservation Act.*
1976	Commonwealth	Mining for minerals sands on Fraser Island (Qld) stopped.
1979	NSW	Terania Creek road blockaded.
	Tasmania	Tasmanian Wilderness Society formed.
1980	Tasmania	South West Conservation Area listed on Register of National Estate by Heritage Commission.
1982	Tasmania	Blockade to save the Franklin River commences.
1982	NSW	Wran Government stops logging in rainforests and proposes World Heritage nomination.
1983	Commonwealth	Labor Party wins federal election. Bob Hawke promises to stop construction of Gordon-below-Franklin dam.
	Commonwealth	*World Heritage Conservation Act* passed and upheld by the High Court. Construction of Franklin River Dam stopped.

Little Desert and the Land Conservation Council in Victoria

In 1968 a scheme to clear and settle the Little Desert area in western Victoria was vigorously promoted by Sir William McDonald, the local member and Minister for Lands, Conservation, and Rural Finance and Settlement.[11] In the long tradition of clearing even poor land and 'horse-trading' Crown land between departments, a powerful Minister could have expected to have his way. But society had changed, and a great controversy erupted in the press as environmentalists claimed the area for a nature reserve and agricultural scientists showed it would be uneconomic anyway. It escalated into a political contest which was reflected within the ruling Liberal Party, and between it and the Country Party which voted with the Labor Party to defeat the scheme.

To avoid such public brawls, Victoria replaced its inter-departmental Land Use Advisory Council with the Land Conservation Council (LCC) in 1970, which was to examine and recommend how all public land in the State should be used. Notably, the LCC was consituted under an independent chairman with not only the permanent heads of eight relevant departments, but also three conservationists.[12] Its establishment opened up critical questions because the permanent reservations of state forests, which had been so slowly built up, could no longer be assured; even less could the continued use of unoccupied Crown lands for timber production be assumed. Moreover, it facilitated critical questioning by incorporating two rounds of public participation in its processes. The legitimacy for such questioning in Victoria provided a precedent for environmentalists in other States.

Lake Pedder and the Franklin River in Tasmania

Proposals by Tasmania's Hydro-Electric Commission to build dams in the south-west of the State which would flood Lake Pedder and later the Franklin River led to major environmental contests which have been widely discussed elsewhere.[13] However, their implications for the forest contests were so important that they warrant a brief review here.

Lake Pedder marked a turning point from the era of coexistence of the national parks movement with conservation as wise use to the era of environmentalism; from gentlemanly persuasion to political activism. The Hydro-Electric Commission met unexpected opposition to its long-established program of building dams when, in 1967, a few Tasmanian conservationists formed a committee to save the lake from being flooded, and the newly formed Australian Conservation Foundation (ACF) pressed the case. The ACF published scientific reports, tried to persuade the State government, and finally in 1972 its prestigious president, Sir Garfield Barwick, wrote a public letter to the Premier.[14] It was to no avail: the Serpentine Dam was built and the lake flooded. Meanwhile and independently of the ACF, *ad hoc* groups had sprung up in the mainland capital cities to make a campaign run by the Save Lake Pedder National Park Committee in Hobart into a major national issue. Whereas Sir Garfield Barwick had written about the needs to preserve the integrity of national parks and assess impacts when planning land-use, the campaign had been one for the existence of a wild and beautiful place.

Although a bitter defeat, Lake Pedder showed that a strategy of campaigning through the media to gain mass support might be more powerful than placing rational arguments before government bodies. Rational arguments, aided by leaks from sympathisers within the bureaucracies, were still needed, but they were best deployed politically. For this, it was critical to build national support for State campaigns, and it was apparent that the Commonwealth had a potential role. So it proved when, in 1977, the Hydro-Electric Commission announced its plan to build the Gordon-below-Franklin dam and flood the spectacular wilderness scenery through which the Franklin River flowed.

The Tasmanian Wilderness Society (TWS) was new and small, with only some 400 members, but it was determined to fight the plan. When Doug Lowe was elected as Premier of a Labor government in 1979, it seemed that the dam would not be built, but the combined forces of elements within the ALP, the Liberal Opposition, and the Hydro-Electric Commission succeeded in toppling him as Premier, followed in 1981 by the election of a pro-development Liberal State government led by Robin Gray and the start of work to dam the Franklin.

An important feeling developed that it was not just a Tasmanian issue and that all Australian people should have a say in the outcome.[15] The popular imagination was fired by the notion of wilderness presented in beautiful images as something sacred.[16] It was intrigued by Bob Brown, the TWS director from 1979, who appeared more and more on the nation's television screens as Tasmanian politics gyrated. He had a personal integrity and a dedication to preserving the wilderness which inspired the environmentalists, and an other-worldliness which many Tasmanians derided, hated even.[17] He seemed created by the media as a myth in his own time. A blockade was organised to prevent work on the dam which attracted 2600 volunteers from all walks of life and parts of Australia, even 67 from other countries. More than 1200 people were arrested and 167 jailed, all of which attracted extensive media coverage. Eventually the national ALP promised to stop the dam as part of its 1983 election campaign; its success was aided by the ACF's endorsement and the Wilderness Society's campaigning in thirteen marginal electorates. The new Prime Minister, Bob Hawke, kept his election promise by passing the *World Heritage Conservation Act* and successfully defending it in the High Court. Construction of the dam was stopped.

A national interest

The sense that there was a rightful national and public interest in the environment which went beyond the interests of particular States emerged between Pedder and the Franklin issues, and found political expression once Gough Whitlam's Labor government took office in 1972. Then the Commonwealth signed the international *Convention for the Protection of the World Cultural and Natural Heritage*, which obliged it to identify and protect sites of world significance, and it set up the Hope inquiry into Australia's heritage which recommended conserving sites of national significance as part of the 'National Estate'.[18] The

Heritage Commission was set up in 1975 to advance this.[19] The process for world and national heritage was similar: sites or 'properties' of various sorts were nominated, placed on an interim register, assessed by experts against detailed criteria, and if found worthy finally listed.

The Commonwealth also passed its environmental impact assessment (EIA) legislation to evaluate development projects in 1974.[20] The process requires a developer to make a Draft Environmental Impact Statement (EIS) describing the proposal and its impacts, a task commonly undertaken by consultants. Public submissions are then received which the developer is supposed to take into account before lodging a Final EIS, which with further submissions is then 'assessed' by the environment department.

Both the heritage and EIA legislation enabled the relevant Commonwealth Minister to be advised about the impacts of a development before deciding whether it should proceed, be modified or be prohibited. But they could not apply directly to the management of resources such as forests, which were constitutionally a State responsibility. Only by finding ways to invoke some of its indirect powers, over trade, foreign affairs, or investment for example, could the Commonwealth assert the national interest. Both processes allow considerable discretion and their findings are not binding on the Minister with the final say who has only to 'take them into account'. All of which ensured that they remained highly politicised.

Land and Environment Court (NSW)

The sense of a rightful public interest in the environment within States led the States to develop their own EIA processes, generally somewhat simpler than the Commonwealth's. However, New South Wales added a specialist Land and Environment Court in 1979 to monitor its EIA process.[21] The court had experts as well as lawyers for its commissioners, adopted less formal procedures than traditional courts, and most notably, gave any member of the public standing to challenge a developer's or agency's compliance with the EIA processes – a provision which enabled environmentalists to challenge the Forestry Commission legally, even though it became progressively harder for them to meet the costs.[22]

Environment movement

The Australian environmental movement expanded, formed various organisations, groupings and alliances, proliferated and splintered into more radical groups in an ever-changing manner. In 1950 there were

Conflicting views on the forest contests expressed on bumper stickers. (John Dargavel)

probably less than 50 conservation societies; by 1970 there were over 200; and by the 1980s a directory listed 800 organisations covering all sorts of areas, interests and activities, but thought there were 300 to 400 more, many of which were probably local and transient.[23]

Like most 'community' organisations, most environmental organisations are run by an often uneasy mix of an elected committee

Table 6.2 Date of formation of conservation and environment non-government organsations concerned with forests

Year	*Base*	*Organisation*
1901	National	Royal Australasian Ornithologists Union
1930	Queensland	National Parks Association of Queensland
1933	NSW	National Parks and Primitive Areas Council
1930s	National	Men of the Trees
1944	Victoria	Natural Resources Conservation League
1948	National	Australian Committee of IUCN
1952	Victoria	Victorian National Parks Association
1955	NSW	Nature Conservation Council
1957	NSW	National Parks Association of NSW
1962	Tasmania	South-west Committee
1962	Queensland	Wildlife Preservation Society of Queensland
1962	SA	Nature Conservation Society of SA
1963	ACT	National Parks Association of the ACT
1965	National	Australian Conservation Foundation
1965	National	Australian Council of National Trusts
1967	WA	Conservation Council of Western Australia
1968	Tasmania	Tasmanian Conservation Trust
1968	NSW, National	Colong Foundation for Wilderness
1969	Victoria	Conservation Council of Victoria, known as Environment Victoria from 1993
1969	NSW	South Coast Conservation Society
1969	Queensland	Queensland Conservation Council and Environment Centre
1970	NSW	National Parks and Wildlife Foundation
1971	SA	Conservation Council of South Australia and Conservation Centre
1972	NSW	Total Environment Centre
1972	Victoria	Geelong Environment Council
1972	Tasmania	Tasmanian Environment Centre
1972/3	SA, Victoria	Friends of the Earth – 1975 NSW
1973	Queensland	Capricorn Conservation Council
1974	National	Australian National Parks Council
1974	ACT, NSW	Canberra and SE Region Environment Centre
1974	NSW	Cole Committee
1974	Tasmania	Launceston Environment Centre
1975	Queensland	Townsville Environment Centre and Townsville Regional Conservation Council which became the North Queensland Conservation Council in 1985
1975	Queensland	Wide Bay Burnett Conservation Council
1975	Queensland	Townsville Regional Conservation Council
1975	National	Ecological Society of Australia
1975	WA	Campaign to Save Native Forests (to late 1980s)
1975	WA	South-West Forests Defence Foundation
1975	WA	Environment Centre of Western Australia
1976	Tasmania	Tasmanian Wilderness Society

Table 6.2 Date of formation of conservation and environment non-government organsations concerned with forests (cont'd)

Year	*Base*	*Organisation*
1976	Victoria	Native Forest Action Council
1976	WA	WA National Parks and Reserves Association
1977	NSW	North Coast Environment Council
1978	National	World Wide Fund for Nature Australia (formerly World Wildlife Fund)
1979	Queensland	Sunshine Coast Environment Council
1979	WA	Men of the Trees (restarted)
1980	Queensland	Men of the Trees (Queensland)
1981	NSW	Rainforest Information Centre
1981	ACT, NSW	Conservation Council of the SE Region and Canberra
1981	Queensland	Cairns and Far North Environment Centre
1982	National	Australian Trust for Conservation Volunteers
1982	Queensland	Rainforest Conservation Society
1983	NSW, Victoria	Albury–Wodonga Environment Centre
1983	NSW	Big Scrub Environment Centre
1983	NT	Environment Centre NT
1984	Victoria	East Gippsland Coalition
1984	NSW	Illawarra Environment Centre
1984	Tasmania	Forest Action Network (FAN)
1984	Queensland	North Queensland Conservation Council
1985	NSW	South-East Forest Alliance (SEFA)
1986	NSW	Coffs Harbour Environment Centre
1987	Victoria	Rainforest Action Group
1987	WA	Denmark Environment Centre
1988	National	Earth First!
1988	NSW	Byron Environment Centre
1989	Queensland	Gold Coast and Hinterland Environment Council
1989	NSW	North-east Forest Alliance (NEFA)
1989	NSW	Clarence Environment Centre
1989	NSW	Nimbin Environment Centre
1989	Queensland	Toowoomba and Region Environment Council and Environment Centre
1989	Tasmania	Deloraine Environment Centre
1990	WA	WA Forest Alliance
1991	NSW	Sutherland Shire Environment Centre
1991	WA	South West Environment Centre
1992	WA	Busselton–Dunsborough Environment Centre
1993	WA	Kimberley Region Environment Centre
1993	Victoria	East Gippsland Forest Alliance

or council, too few paid staff, and voluntary helpers. Their members have very varied levels of personal commitment, from simply subscribing to dedicating virtually their whole lives to demonstrations, running organisations, or leading peak national bodies. As with other groups with high ideals, passions can run high, schisms develop, and making decisions can be laborious and frustrating, especially in groups which reject hierarchical structures in favour of cooperatives. Networks of personal contacts formed by friendships, newsletters, telephone 'trees', overlapping memberships in organisations, and shared ideals are an essential feature of the movement. Many of the leaders in the peak national and state bodies keep in frequent personal contact, especially during intense negotiations with governments, although this can create tensions between what can be seen as an elite leadership and the mass of environmentalists.[24]

Individuals take very different passages into direct action protests in the forests. To some, 'saving' the forests is a life-time commitment. Others come as members of environmental groups, joining the protests during their holidays from work or college. Some just drift along with friends without much prior knowledge of the forest but become more dedicated as they experience its threatened tranquillity. And to some, forest protest becomes a style for living, for participation, something to be experienced, enjoyed.

Some of the organisations involved in the forest contests are described in what follows, from national organisations, through State and regional bodies to local and informal groups.

National organisations

Australian Conservation Foundation[25]

The Australian Conservation Foundation (ACF) is the major national body which speaks and campaigns across a range of environmental issues, of which forest use has always been prominent. It was started in 1965 by forward-looking scientists, bureaucrats and well-connected figures of the Melbourne establishment, following a suggestion by Prince Phillip, the Duke of Edinburgh, who had an abiding interest in the work of the World Wildlife Fund. Its first president was the Chief Justice of the High Court, Sir Garfield Barwick, and its patron was the Governor-General.[26] With such connections, a government grant of $20 000 was readily obtained, funds raised, an executive officer appointed, and members recruited; by 1970 there were 3500 individual members, and 350 conservation organisations and companies. At first the ACF worked by scientific investigation, public education and persuasion. Its objective was 'always to try to make positive suggestions which will influence government policy in a constructive way', and it emphasised the 'education of the younger generation in conservation'.[27] To make its persuasion more effective in the corridors of power, it hired a company director, expert in public relations, as its executive director.

Environmental problems piled up which such gentlemanly persuasion, tried over Lake Pedder, could not stop. At its next annual general meeting a group of energetic environmentalists replaced the existing council and promptly dumped their prestigious executive director in favour of an ardent environmental advocate, Geoff Mosley.[28] Although a few conservatives resigned, membership grew and the Commonwealth's grant was increased to $150 000 a year.

The changed ACF entered into campaigns and projects with a will, concentrating on forestry, wilderness, and energy and natural resources. It continued to make submissions to the ever increasing number of State and Commonwealth inquiries, it held conferences, and it published extensively. Membership continued to increase through the 1980s and although Victoria remained the ACF's base, it set up State offices where it could, and eventually formed thirty branches of members around Australia. With increased funds, it employed project officers to work on issues such as forests, south-west Tasmania and the wet tropics.

As the importance of the Commonwealth in environmental issues grew, so did the role of the ACF as a national peak body that could deal with governments. It set up a national liaison officer in Canberra to lobby politicians and bureaucrats, and in 1986 appointed Phillip Toyne as director. Articulate and forceful, he became an almost nightly figure on Australia's television screens as the forest contests raged on.[29] In some ways, the ACF returned to the mode of its founders as it negotiated in the corridors of federal power, but the persuasiveness of its argument was enhanced by applying political pressure to a degree that Sir Garfield Barwick would never have countenanced.

Membership continued to increase, and various fund-raising and marketing schemes, including that epitome of consumerism, catalogue selling – albeit of 'environmentally friendly' products – was introduced to raise money.[30] However, its bureaucratic structure was more expensive than its income could support, so that it was forced to close some of its State branch offices in 1993.

Wilderness Society[31]

The Wilderness Society evolved as a national body to campaign for the preservation of wilderness after the success of its campaign to save the Franklin. It started 1976 after the Lake Pedder campaign failed when, rather than overthrow the moderate Tasmanian Conservation Trust, the Tasmanian activists first formed the radical, ecocentric South-west

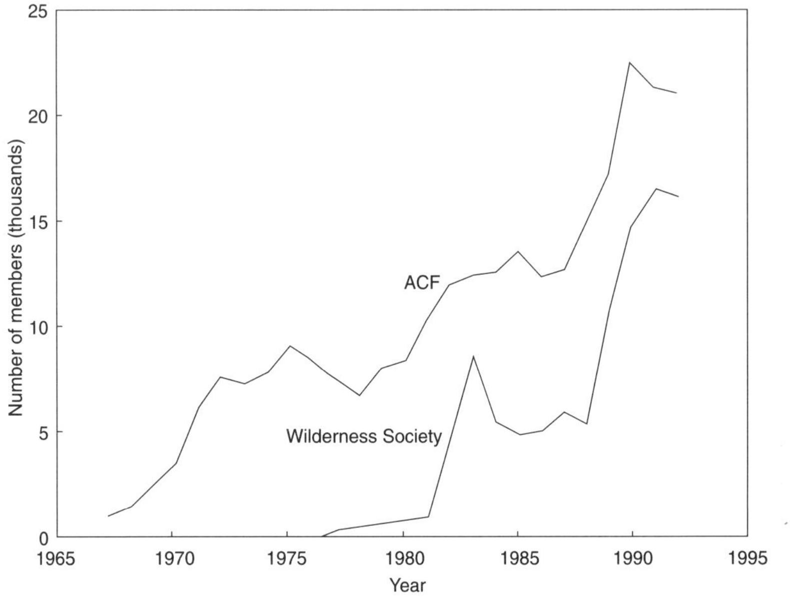

Figure 6.1 Membership, Australian Conservation Foundation and Wilderness Society, 1967–92

Source: ACF, *Annual Reports*; Wilderness Society, pers. comm.; Holloway 1986; Green 1981.

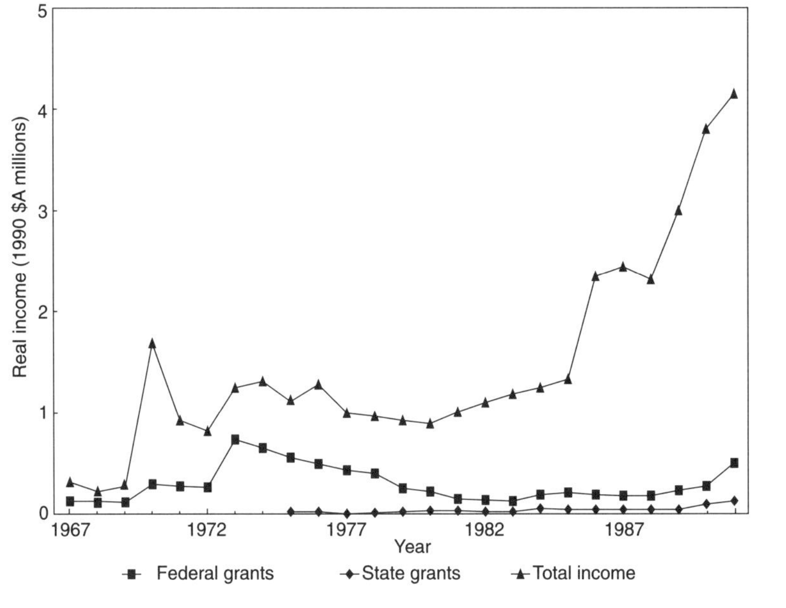

Figure 6.2 Real income, Australian Conservation Foundation, 1967–92

Wilderness, No Compromise.

A poster of Farmhouse Creek produced as part of the Wilderness Society's campaign over the Lemonthyme and Southern Forests, Tasmania.
(Rob Blakers, The Wilderness Society)

Tasmania Action Committee, and then joined with the United Tasmania Party to form the Tasmanian Wilderness Society (TWS).

The Franklin campaign rapidly turned the TWS into a national organisation with seventy branches around Australia, shopfronts in the capital cities, a flow of donations and an increasing membership. It also put the preservation of wilderness on the mainstream political agenda in a way that it never had been before. Afterwards, the Tasmanian Wilderness Society became The Wilderness Society and widened its interests to other States. It kept its base in Tasmania but, like the ACF, was able to increase its membership and engage a national liaison officer in Canberra to exert its pressure in the centre of Commonwealth power. It too took a more commercial approach to raising funds by upgrading its shops and dressing people as cute koalas to pester the public for donations.

Branches of international agencies

Some of the international conservation and environment bodies have branches in Australia which take an interest in forest issues. Most important is the semi-governmental International Union for the Conservation of Nature and Natural Resources – now the World Conservation Union (IUCN) – which has an Australian committee. IUCN evaluates nominations of sites to be listed as being of significance as World Heritage. It was IUCN's listing of areas as World Heritage that gave the Commonwealth the power (under its constitutional obligation to comply with international conventions to which it was a signatory) to over-ride Tasmania over the Franklin River, and Queensland over its wet tropical rainforests.

The World Wildlife Fund (WWF) – now World Wide Fund for Nature – formed an Australian affiliate in 1978. It funds scientific research and urges government agencies to take measures to preserve endangered animals. Greenpeace has an Australian branch which has campaigned strongly against pollution from pulp mills.

State level organisations

National Parks Associations

Associations to lobby for the creation and protection of national parks are some of Australia's oldest environmental organisations. They also arrange excursions, bushwalks, and evening lectures for their members, publish newsletters, and arrange displays as part of their general public advocacy of parks. Two examples of the associations are described in this section.

The National Parks Association of Queensland was founded in 1930 to preserve the parks that had already been declared and to argue for more. It worked on the principle of 'stick to the facts, no exaggerations, no hysterics and no histrionics' in persuading government departments and politicians to add, piece by piece, to the number of parks.[32] It gradually built up its membership to over 1300 by 1979. In its modest way, the Queensland Association helped both the NSW and Western Australian Associations to get started.

The Victorian National Parks Association was not formed until 1952. It is more outspoken and slightly larger, with over 3000 members, than its Queensland counterpart. It fought the Little Desert affair almost single-handedly, lobbied successfully for the *Victorian National Parks Act*

1975 and was particularly active in advocating a large national park in the Victorian alps.[33] By the 1980s its president was calling for the organisation to be more active. With much of the membership being 'well-educated middle class' which provided 'age and conservatism for balance' it was thought that the association needed 'the hard cutting edge and raw anger of youth to achieve results'.[34] By the mid-1980s more radical people had been elected to VNPA's council and it was undertaking explicitly political research into public attitudes. Its slightly more activist stance was expressed by entering into alliances such as the East Gippsland Coalition, mentioned later.

National Trusts

National Trusts were established in each State and Territory between 1945 and 1976, on eminently British lines, to acquire and conserve historic buildings. They are large, conservative, well-connected and well-supported organisations with a federal council, more than 75 000 members, and an annual budget for 1991–92 of over $6 million.[35] The heritage of landscape, long a British concern, became of more interest in Australia after the Hope inquiry emphasised that it was as much part of the national heritage as historic buildings.[36] The Trusts became actively involved in nominating forest areas for inclusion on the Register of the National Estate, making submissions to public inquiries, and lobbying governments to promote heritage legislation. Some of them now acquire and manage bushland and other reserves, and a special one was initiated by Bob Brown to buy forest land. A National Parks and Wildlife Foundation was started in New South Wales in 1970 which raises funds largely from commercial sources for, among other things, buying land to add to the national parks.

Conservation councils and environment centres

Two sorts of related organisations evolved to cover the varied environmental interests and groups in the States and particular regions. Conservation Councils were created as bodies to bring together various groups in their areas and provide a voice which would be better recognised by State governments. Most receive a small government grant which helps them maintain offices and hire coordinators or directors. They investigate problems, make submissions to inquiries, lobby politicians, and conduct campaigns on various issues. The Nature Conservation Council in New South Wales was the first to be created, in 1955. It acts as the 'umbrella' organisation for about ninety environmental, scientific and recreational organisations, and operates the Environment Centre of NSW in Sydney. Regional councils were formed from the 1970s and continue to expand. For example, the North Coast Environment Council, formed by nine groups in 1977, grew to have forty-one member bodies and kept forest questions on the top of its agenda.[37]

The twenty or so environment centres around Australia are both shopfronts for the environmental movement and resources of detailed information. They often run education and community awareness campaigns, and provide meeting places for environmentalists. Many were set up by conservation councils and shares premises with other groups. There are two in Sydney: the Environment Centre of NSW, which functions much like those in other States; and the Total Environment Centre, which is at the more radical and active end of the spectrum of environmentalism and, under the leadership of Milo Dunphy, has been particularly concerned with forests and national parks.

Forest Alliances

Alliances emerged around specific issues or campaigns as another sort of grouping. Four were notable in the forest contests: the Forest Action Network in Tasmania, the East Gippsland Coalition in Victoria, and the South-East and North-East Forest Alliances in New South Wales. They had different structures and modes of working.

Victoria

A loose coalition of about ten environment groups campaigned successfully in 1975 stop a proposal for woodchipping in East Gippsland. They realised that an organisation to focus specifically on native forests was needed. The resulting Native Forests Action Council (NFAC) was formed in 1976. It launched a 'Forests Campaign' in 1983, but this became overshadowed by the urgency of supporting the campaign to save the Franklin River. Renewed proposals to build a pulp mill in East Gippsland led NFAC to launch an East Gippsland campaign in 1983 with financial assistance from the ACF. However the NFAC only had about 800 members and needed to form an alliance with other organisations concerned with the issue to make a coordinated and effective campaign.

The East Gippsland Coalition (EGC) was formed in 1984 by State and national organisations to conduct a campaign for more parks to be reserved in the region, and for sound management of those forests used for timber production.[38] Victoria's National Parks Association,

Conservation Council, and Native Forests Action Council were joined by the ACF and TWS in establishing the coalition. Notably, key activists employed by these bodies worked under the banner of the EGC when they spoke on East Gippsland issues. The EGC was able to hire three coordinators to produce reports and submissions, lobby politicians and bureaucrats, publicise the issue especially with the media, work with many active volunteers, and raise funds. It recruited some 3000 members and supporters to become larger than any of its State partners and acted in its own right from 1987 onwards. Its campaign ended and the Coalition lapsed, to be replaced in 1993 by the East Gippsland Alliance made up of the Concerned Residents of East Gippsland, TWS and Friends of the Earth.

Tasmania and New South Wales

The Forest Action Network (FAN) was formed in Tasmania in 1984 to limit and control woodchipping in response to the EIA process for licence renewal. It drew the Tasmanian Conservation Trust, ACF, TWS and local groups together for the purpose.[39]

The South-East Forest Alliance (SEFA) was formed similarly in New South Wales in 1985 to draw various environmental groups and networks together at a time when the licence for Harris–Daishowa's woodchip operation out of Eden was due for renewal and an EIA was held.[40] It brought together national, State, regional and local organisations comprising: ACF, TWS, Fund for Animals, Total Environment Centre, Nature Conservation Council, Conservation Council of the Canberra and South-East Region, Far South Coast Environment Group and the Towamba Catchment Protection Association. Two other local groups, the Tantawanglo Catchment Protection Association and the Yowaka–Egan Peaks Group, joined later.

The North-East Forest Alliance (NEFA) was formed in 1989 to campaign for the protection of old-growth forests, rainforests, endangered species and water quality in northern New South Wales. It attempts to compel the Forestry Commission to manage the state forests in what it believes to be an ecologically sustainable manner.[41]

NEFA operates as a coordinating network of organisations and individuals without a constitution, formal organisational document or list of members. Almost all environmental organisations in the region, and some national and State organisations, such as the NSW National Parks Association for example, are affiliated with it in some way. NEFA has five loosely defined regions with two voluntary coordinators in each. Individuals, who are generally members of other groups, contribute their skills, time or money as they can. Regular meetings are open to anyone interested. This open, informal, amorphous structure is deliberate in its rejection of hierarchies and offers some advantages in mounting legal challenges against the Forestry Commission.

NEFA is able to operate in an efficient and highly sophisticated manner at times. For example, its volunteers set up a radio-telephone network to provide protesters in the forest with more up-to-date information on the progress of court cases than the police network. It relied mainly on fund-raising by local groups and donations, some quite large, and many received through ACF. Prior to the 1993 Budget, the latter route enabled donors to secure a tax concession. The vigour of its campaigns made it the most prominent environmental organisation in the region.

Local groups

There is not space here to describe the hundreds of local groups concerned about Australia's forests. What should be noted, however, is that there are many environmentalists and small groups in rural areas, and that they are formed from concerned long-term residents, such as farmers in the catchment associations, as well as new settlers. Notably, they battle on with environmental concerns in their areas, dealing with local foresters and officials as best they can, while much of the real power is exercised in the capital cities.

Informal and anarchist groups

At quite the opposite extreme from the national leadership were people protesting in the forests outside the organisations so far described, many of whom saw direct action as a first, not last resort and were not necessarily committed to non-violent means. One set of about five to seven people, operating under the name of the Nomadic Action Group, moved from protest to protest around the country, from wilderness issues in south-west Tasmania, to strikes in Roxby Downs, to forest issues in Victoria and south-east New South Wales, gathering new supporters in each area. Under the name of the Forest People, they were involved in a blockade in Errinundra in Victoria.[42] Although in the short-term they attracted considerable media attention and prevented logging in the area of their blockade, they severely damaged the long-term campaign of local groups and the East Gippsland Coalition there, and completely wrecked any tentative relationship with the timber workers. A label of The Ferals is used by some people involved in forest protests in northern New South Wales, but is also applied to people involved in a hard drug culture there.

Scientific understanding

Environmentalism called for a much wider scientific understanding of the natural world and the impacts of human patterning. Instead of the 'reductionist' understanding, gained conventionally by accumulating detailed studies of particular components of the environment, ecophilosophers advocated a 'holistic' understanding of whole ecosystems. To this end, systems theorists helped ecologists model the energy, water and nutrient flows, and the inter-relationships between the various plants and animals in the forest. Ecology flourished, aided in Australia from 1975 by an Ecological Society with 900 members and its own journal. The demand for information, research and education grew, it seemed, exponentially. The environmentalists quizzed the agencies, the agencies needed information and staff to advocate or rebut change, and many people, especially in the new middle class, wanted to study the issues for themselves. People, budgets, societies, research laboratories, schools, colleges, universities, and organisations of all sorts were drawn in to meeting the demand, or giving their own twist to it.

The Commonwealth set up the Australian Biological Resources Council in 1973 to fund and integrate taxonomic and biogeographical surveys. While such patient work slowly expanded knowledge, it was the survey of the major plant alliances, led by Professor Specht and conducted nation-wide for seven years, which provided a 'scientific basis for conservation at a level that has never been previously possible'.[43] Almost half the alliances were found not to be adequately protected in existing reserves, and with the imprimatur of the Australian Academy of Science, calls were made for establishing a system of ecological reserves 'as a matter of urgency'.[44] Although the Specht report covered only the vegetation, reservation on a sufficiently large scale would, it was thought, ensure the habitat for animals and hence their conservation. But while plant species could be expected to be conserved in the forests, as long as they were regenerated after logging, the same was not true for many animals and birds, especially those that depended on big old trees. Finding the animals and birds, tracing their life-cycles and determining their habitat requirements was a far more difficult and long-term task, especially as it was the rare species that were of most interest. As well, the effects on soils, water, recreation and scenery all had to be added to the research agenda.

It was clear that the native forests needed to be understood not just as resources of timber and water, but as environments. Although the forest services had the forests well mapped by the 1970s, some had only rudimentary knowledge of their timber resources, let alone the soils,

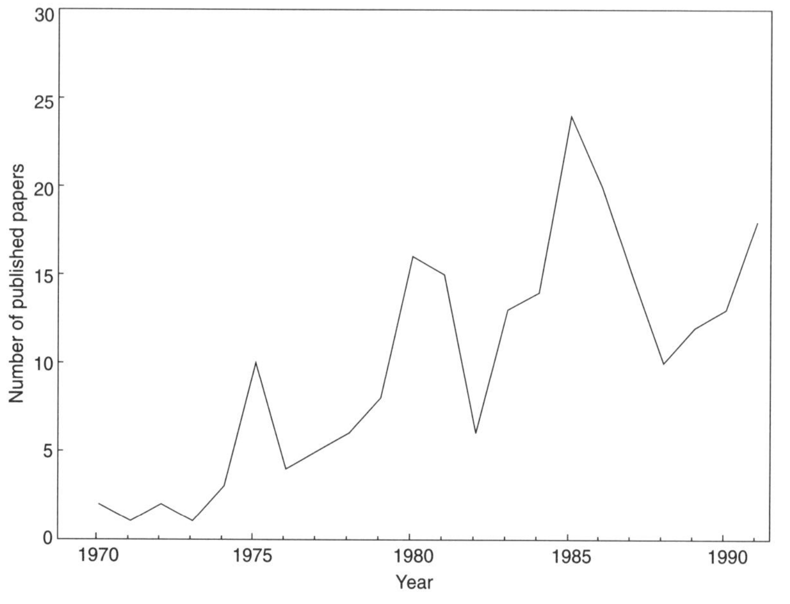

Figure 6.3 Research on the ecological impacts of forest use
Source: RAC (1993).

aquifers, plants, animals, heritage values and all the forest functions which were suddenly of interest. A series of government inquiries in the 1970s found that 'the extreme lack of knowledge on the biological sphere ... is hampering responsible decision making'.[45] Although, scientific understanding of the ecological impacts of logging did slowly accumulate as more research was funded and eventually completed (Figure 6.3), a survey in 1993 concluded that 'the level of information on impacts appears insufficient for most current uses'.[46]

Environmental consciousness and the public mind

Environmental consciousness diffused from the environmental movement and its centres to enter the public mind and make the environment one of the major issues of the times. The extent to which it did so was recorded in many public opinion surveys and the Australian Election Studies conducted after federal elections. The intricate details and differences of these surveys have been analysed elsewhere, and only the principal features strongly established by the 1990s and relevant to the forest issues need be noted here.[47] Unemployment was of the greatest concern to Australians, but almost everybody was also concerned about the environment. Most people thought that the government should do much more to protect it, even if industries might have to raise prices or the government increase taxes. The attitudes were surprisingly uniform, mostly with barely measurable differences across the conventional divides of class, occupation, education and gender. Age sometimes stood out, particularly so over logging the forests, which two-thirds of young people (18–24 years), but only one-third of older people (55 and over), thought was 'very urgent'.[48]

The most comprehensive survey of public attitudes to forest issues was conducted for the Resource Assessment Commission in 1991 and reflected many of the general attitudes.[49] Most people (75 per cent) saw the preservation of the forests as important, even if timber workers lost their jobs (67 per cent) or if it cost more to do so (53 per cent). They believed that clear-felling harms native wildlife and plants and that it was being done too quickly (67 per cent). They saw the forests as important for a whole range of values, including employment and the economic benefits (Table 6.3). Two sets of attitudes – concerning preservation and concerning exploitation of the forests – were held somewhat separately; some people (38 per cent) consistently supported

Table 6.3 Importance of forest management factors

Factor	*Proportion of responses in 'very important' and 'somewhat important' categories*		
	National (%)	*Eden–Monaro Gippsland* (%)	*Difference* (%)
Protect water quality	98	98	0
Prevent soil erosion	97	97	0
Protect wildlife and plants	97	96	–1
Counter Greenhouse effect	93	90	–3
Provide wilderness	85	80	–5
Provide employment	79	87	+8
Allow large old trees to grow	78	69	–9
Reduce imports of timber	75	98	+23
Areas for bushwalking/camping	72	68	–4
Maintain forest industries	72	81	+9
Attract tourists	64	61	–3
Produce house and furniture timber	63	77	+14
Protect Aboriginal values	59	56	–3
Supply timber for export	44	36	–8
Provide export woodchips	33	42	+9

Source: Social Science Data Archive (1992).

a single view, a few (8 per cent) both concurrently, but most people (54 per cent) took a middle position on at least one of the various topics investigated.

A very specific question was asked about logging in National Estate forests – a sore topic at the time – and people were asked to score their attitudes on a scale between banning logging even if it meant loss of jobs and export income at one extreme, and allowing logging to create more jobs and export income at the other (Figure 6.4). Opinion was divided, with slightly more people in favour of banning logging than allowing it, but with many (23 per cent) in the middle or within a one-point score of it (a further 32 per cent). Many of these opinions may not have been well informed because slightly over half the people (53 per cent) believed incorrectly that National Estate forests had protection equal to or greater than that in national parks.[50]

The forest survey included a more intensive sampling of people in the agricultural and forest region of the Eden–Monaro and (East) Gippsland electorates. Generally, people in this rural region held attitudes very similar to those held nationally. Not surprisingly, they were slightly more conservative on social and political questions, less concerned about noise, pollution and lack of access to open space, and more concerned about land degradation. On the specific forest issues, the people in the region attached high levels of importance to the environmental factors, though slightly less than people did nationally. The substantial difference was over the greater support given to logging in National Estate forests.

The shape of environmental consciousness in the public mind, revealed by these surveys, helps explain how politicised the forest

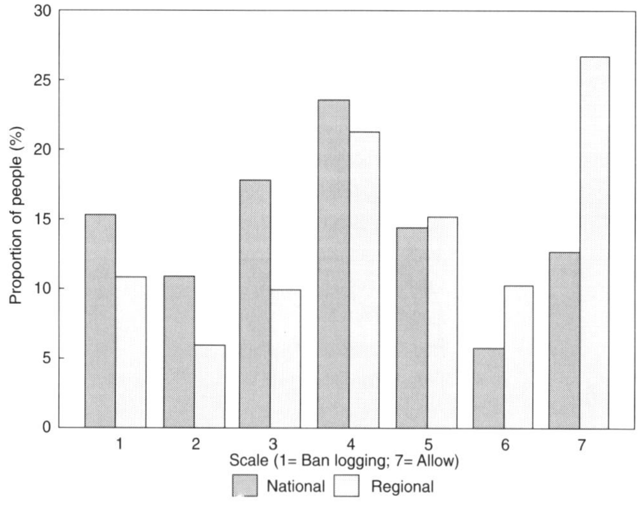

Figure 6.4 Attitudes to logging in National Estate forests
Source: Social Science Data Archive (1992).

contests became. Almost everybody had come to feel that forest issues were important and, although significant numbers appeared to be on one 'side' or the other, the majority held middle views, believing that environmental and economic issues were equally important. Thus, influencing this middle ground through education and especially the media was critical for the contestants.[51]

Media[52]

Learning from its Terania Creek and Franklin campaigns, the environmental movement became very good at gaining media attention; blockades, barricades, tripods, tree sitters, protests, banners and arrests had all the elements of drama and display which made for good television; the movement made sure that they did. Posters and the popular wilderness calendars readily portrayed raw logging sites as hideous and the natural forests as beauteous, while the increasing number of nature documentaries on television aided the cause effortlessly. In print it did well too, with letters to newspapers, pamphlets, bumper stickers, and so forth. In all it had mass and multiple impacts.

The industries could not match it in the popular mind; they had the money but not the numbers and were late in the day. The Forest Industries Campaign Association (FICA) launched a vigorous anti-environmentalist campaign of television and press advertising in 1986, but this confrontationist campaign was found to have had no lasting impact and FICA switched to a more positive approach.[53] Industry lobbyists also commissioned studies, wrote frequently to newspapers, issued copious press statements, and distributed pamphlets. In milder vein, many companies presented their image and wood products as being 'environmentally friendly' in their annual reports, pamphlets and advertising.

The capital city newspapers presented both the environmental and industrial views with what seems to be roughly similar coverage.[54] In the news columns, much depended on which side had a story to tell at the time, because reporters had little of their own research to guide them. The editorial columns and leading articles often supported environmental positions, saying that the native forests were threatened by the 'scourge of woodchipping' for example.[55] The major papers also gave industry lobbyists columns in which to express their case,[56] while the *Mercury* in Hobart and some local papers in the forest regions, such as the *Snowy River Mail* in Orbost, generally took a pro-industry stance. An analysis of newspaper articles on woodchipping in New South Wales from 1969 to 1985 found that in only two years did the number of critical items exceed the number of pro-industry items.[57] Television news, being geared to very short segments, gave little background information and often concentrated its images on the most extreme aspects of logging or the most outlandish protesters, which neither side liked. Television documentary programs, like *Four Corners*, were accused of undue bias by foresters and industrialists, while a publicity film put out by the Western Australian forest service was similarly accused by environmentalists. Overall, rather more people (39 per cent) thought that television was biased in favour of the 'greens' than thought it was not (25 per cent).[58]

Even though the statements of both sides were well covered in the media, the public gave those by the large companies and industry groups little credence. Survey results showed that: science and nature programs on television, the CSIRO, and educational institutions were believed by three-quarters of the people; environmental groups by half the people; television news, current affairs programs, and government statements by one-third of the people; newspaper articles by one-quarter; and large companies and industry groups by only one-tenth.[59] The industrialists fared badly in the media because the frequent exposure and intensity of the forest conflicts built their importance in the public mind, and because of the low regard with which Australians hold governments and companies. But building importance did not lead, as the moderate conservationists had originally hoped, to much change in the way the forests were used. For substantial change, the issues had to be contested in the political and bureaucratic arenas.

CONTESTING

FROM THE 1970s

Wood production should be transferred from native forests to areas of new forests and plantations established outside the current forest estate.

Australian Conservation Foundation (1983), Forestry in Australia, Policy Statement No. 36

This industry and the jobs of its employees are under threat from an orchestrated campaign by a vocal minority of extreme environmentalists who have singled us out as a whipping boy for their own selfish purposes.

Thorry Gunnersen, Chairman of the Forest Industries Campaign Association, at the launch of a public awareness campaign, 1987

Many trade unionists and traditional ALP supporters believe that ALP Governments have disregarded them in the interest of wooing the so-called 'greenie' support.

Tasmanian Trades and Labour Council, The ALP – *conservation with development ... a growing awareness*, brochure produced for the 1988 ALP Conference.

The contests over how Australia's forests and plantations should be fashioned, by whom and for what purpose, raged through the 1970s, 1980s and early 1990s. They were more complex than the polarisations of the media or the political processes could readily admit. Although the themes are interwoven, Chapter 7 concentrates on the ecology of the forests and the existence of species, Chapter 8 on economic and regional issues, and Chapter 9 on political and bureaucratic processes.

CHAPTER SEVEN

CONTESTING EXISTENCE, ECOLOGY AND EXPERIENCE

It was the environmental movement, not capital, labour, the economy, governments or regions, which set the public agenda of forest issues from the 1970s. Not only did it manage to put its own claims for national parks, wilderness, heritage, rainforest, and endangered species high on the agenda, it turned the established use of the forests into a public issue. What had been commonplace became contested. And the issues hardened as the movement diversified and became more radical. Moderate environmentalists sought to ameliorate the impacts by parley; radicals assaulted the social acceptance of that very normality, undermined the legitimacy of the resource regimes, and asserted fundamentalist claims not amenable to parley. What started as debates became fierce contests in the arenas of public opinion, the media, science, and in the state through inquiries, impact assessments, bureaucracies, parliaments, and courts.

This chapter deals with the environmental claims roughly in the order of their appearance on the public agenda. The first part deals with the two earliest and urgent claims to save the forests from felling for woodchip exports and clearing for pine plantations. The next part deals with claims over land use for the expansion of national parks and the preservation of natural heritage, especially as wilderness. The last part deals with claims for the preservation of rainforests, rare species, and all the rich diversity of the forests.

Clear-felling for woodchip exports

The rapid expansion of woodchip exports on a large scale depended mainly on clear-felling the forests in the integrated resource regime. Within five years more than 25 000 hectares were being cut in Australia each year. The area increased subsequently, probably to about 35 000 hectares a year, so that by the 1990s probably about half a million hectares had been cut.[1] The scenes created by the early, hastily arranged logging were hideous; the environmental impacts were substantial, and in a world becoming more environmentally conscious, quickly opposed. Within two years of operations starting at Eden, Richard and Val Routley were writing their angry critique, *The fight for the forests*, which, when it was published in 1973, was read widely.[2] More than anything else, it moved environmental issues from the scientific and official domain to the public agenda. It marked the start of the forest contests.

To the forest services and industries, *The fight for the forests* was a startling document because it attacked not only the environmental consequences of exporting woodchips and planting pines, but the long-held values on which they were based. Suddenly, forestry was being reviled as 'wood production ideology' and the forest services depicted as the 'main destroyers of the remaining natural forest areas', paternalistic, secretive from the public, tied to the interests of the wood industries, lacking relevant expertise in ecological matters, and not capable of giving proper weight to values other than wood production.[3] The Routleys' attack was bitterly resented by the foresters, who felt that they had worked hard to do what had been long expected of them. The Routleys went on and on through detail after detail, but it was their photographs of raw logging sites which did most to convince the public.

The Australian Conservation Foundation (ACF) quickly followed with a pamphlet, *The great forest sell-out*, a group of academics at the University of Tasmania produced *The vanishing forests?* and the Campaign to Save Native Forests (WA) produced *The threat to the forests*.[4] They were accompanied by letters to newspapers, public meetings and pressure on politicians. The objections ranged from impacts on the natural environment to social, economic and political matters (Table 7.1). However, as a Senate inquiry was to note in 1977:

> Formulation of an objective opinion on the environmental aspects of the industry is hampered by a general lack of research knowledge on the nature of Australian forests, their soils, streams, fauna and flora. Meaningful dialogue on the issue has in consequence become increasingly difficult, with the protagonists of both sides tending to become entrenched in their respective positions, and with much of the argument characterised by poorly supported assertions and emotion rather than logic.[5]

In any case, it was clear that Australian forestry was changed for ever: values other than wood production and development had to be taken into account; a great number of people wanted a say in how the forests were used; and 'woodchipping' had become an icon to be vigorously

Table 7.1 Concerns about woodchipping

Topic	*Concerns*
Environmental	
Soil	Declining fertility due to removal of nutrients and erosion.
Water	Siltation of water courses, possible eutrophication due to increase in nutrients.
	Depletion of value as water catchment and increased risk of flooding.
	Increased turbidity.
	Increased salinity.
Flora	Depletion of species, by replacement of non-commercial species, elimination of special habitats, introduction of exotic species.
Fauna	Depletion of species, including smaller less conspicuous ones.
Disease	Increased risk of fungal pathogens such as *Phytophora cinnamoni.*
Pests	Increased risk of pests in uniform stands; consequent risks if insecticides or 1080 poison used.
Fire	Increased risk of fire in young stands; they are able to regenerate naturally.
	Escapes of fire and air pollution from burning cut-over·areas.
Scenery	Loss of value of undeveloped environment.
Wilderness	
Education	Loss of undisturbed environment for outdoor education in independence and survival skills.
Science	Loss of study areas and gene pool.
Resource	
Sawlogs	Over-exploitation threatens long-term future of sawmilling industry.
Waste	Waste wood left in harvesting.
Regeneration	Not all cut-over areas regenerated.
Social	
Recreation and tourism	Loss of value due to loss of scenery.
Restructuring	Shift from small sawmills in rural towns to large mills in main towns.
Economic	
Assessment	Lack of economic assessment before schemes commenced.
Infrastructure	Road tolls insufficient to cover costs.
Royalties	Royalties paid to forest services do not cover costs of regenerating stands.
	Payments to private owners not adequate to cover costs of regeneration, leading to deforestation.
Other industries	Reduction of honey and wildflower production. Reduction of potential for pharmaceutical industry.
Public investment	Public funding of woodchip company through Australian Industries Development Corporation.
Political	
Decision	Forests should be managed in the public interest and choice of how to do so should not be left to the forest services and industries.

Source: ACF (1975); Jones (1975); Campaign to Save Native Forests (1975).

opposed, independently of the supply to the domestic pulp mills by the same integrated regime.

Governments tried to deal with the situation by holding inquiries and conducting impact assessments, while the forest services tried variously to rebut the criticism, improve their public relations, research non-timber aspects of the forests, alter the pattern and ameliorate the impacts of logging, and sometimes just 'turtle along' as if nothing had changed.

Inquiries

The Commonwealth government's first response to the growing opposition to woodchipping was to set up an inter-agency working group. It produced a small report attaching papers about which the group could not agree.[6] One on the biological and ecological aspects found that the consequences of woodchipping were significant but difficult to predict, and the reserves in the regions too small.[7] Another on forest management looked at how woodchip exports could be maintained either by altering practices to reduce their impact, or by intensifying them, possibly by converting some of the land to pine plantations, in order to reserve greater areas for non-wood values.[8] The rift was clear. On one side, woodchipping should be stopped, at least until it was proved environmentally and economically sound; on the other, it should be accepted and managed.

Meanwhile, the Senate had started its own thorough inquiry, which attracted over 250 submissions across the range of interests. Although it found that 'curtailment of the existing woodchip industry program is not warranted on environmental grounds at this time [1977]', it recommended that no new projects should be started, nor licences for existing projects increased or renewed, until the environmental problems were resolved.[9] It produced a wealth of other proposals, which it hoped would be accepted by the States 'in a spirit of constructive co-operation'.

Tasmania, the largest exporter of woodchips, examined the rapidly changing condition of about 900 000 hectares of its privately-owned forests which were within economic grasp of the new woodchip mills. The hard-pressed land-owners scrambled at the chance to sell their previously useless timber so fast that they provided over 40 per cent of the wood exported during the 1970s.[10] Scenes of apparent devastation were thrust before the public as many cut-over areas were burnt and left for rough grazing rather than being regenerated to a new forest. A fear that this would eventually damage the Tasmania's forest economy induced the government to appoint a Board of Inquiry, under Mr (later Justice) Everett.[11] Its recommendations to provide land-owners with advice and financial incentives to regenerate their forests were accepted; its recommendations to raise royalty rates and control the amount cut to a sustainable level were not; and its recommendation

Table 7.2 Public inquiries and Environmental Impact Statements on woodchipping

Year	*Level*	*Inquiry or statement*
1974	Australia	Committee of Inquiry into the National Estate. Included some consideration of woodchipping in its *Report on the National Estate.*
1975	Australia	Working Group on the Economic Economic and Environmental Aspects of the Export Hardwood Woodchip Industry.
	NSW	State Pollution Control Commission. Environmental inquiry into proposed woodchip exports from the north coast of New South Wales.
1977	Australia	Senate Standing Committee on Science and the Environment. *Woodchips and the environment* report.
	Tasmania	Board of Inquiry into Private Forestry Development in Tasmania (Everett inquiry).
	NSW	Harris–Daishowa (Australia) Pty Ltd produced the first EIS for the Eden export woodchip operation.
1978–80	WA	Steering Committee on Research into the Effects of the Woodchip Industry on Water Resources in South Western Australia. Two reports.
1982	Victoria	Interdepartmental Task Force. Reported on the effects of pulpwood harvesting for woodchips in the Otways.
1985	Tasmania	Select Committee of Legislative Council. Included woodchipping in its State forestry report.
	Tasmania	The Tasmanian Woodchip Export Study Group (the Forestry Commission and the exporters) produced the first EIS for Tasmanian woodchip exports.
1986	NSW	Harris–Daishowa (Australia) Pty Ltd produced the second EIS for the Eden export woodchip operation.
1990	Australia and NSW	Joint Scientific Committee on South-East Forests reported on biological conservation in the Eden region.

Source: Adapted from RAC (1992), vol. 2B, Appendix W.

that logging standards should be set to minimise environmental damage had to wait a dozen years before it was introduced.

Environmental Impact Assessments

Two early assessments[12]

The Commonwealth had required woodchip exports to be licensed from their beginning, but from 1975 it was able to require new or expanding mills to comply with its EIA process. The first two proposals, which came from New South Wales, had little difficulty in doing so. On the north coast, Sawmillers Exports engaged a management consulting company, W. D. Scott, to prepare its EIS.[13] As its chips were to come from sawmill waste and thinnings, no additional clear-felling was proposed. Hence, the potential impacts in the forests were thought by the consultants to be limited to visual effects and a possible loss of nutrients, a matter suggested for research. They found it 'difficult to understand the depth of feeling associated with opposition to woodchipping', which they attributed to confusion between 'clear-felling' and 'woodchipping' on the part of an 'anti-woodchip faction'.[14]

On the south coast, Harris–Daishowa wanted to increase the quantity it could export from Eden from 636 000 to 850 000 tonnes a year. The Commonwealth granted an interim rise to 806 000 tonnes in 1977 but required an EIA. W. D. Scott was again engaged.[15] Its statement emphasised that the Forestry Commission, not the company, was responsible for logging standards and that these had been changed to 'lessen the environmental impact'. It cited research by the Australian Museum to claim that 'there has been no adverse effect on the environment'.[16] The reports for both areas were brief but, in the absence of substantial objection, were sufficient to secure Commonwealth approval.

A decade later, the situation had quite changed: the renewal of the five major licences was being opposed not only by far larger and more effective environmental groups, but also within the environmental policy committees of the Australian Labor Party (ALP). They persuaded the party's national conference in 1985 that EIAs should be conducted before any of the licences were renewed. The resulting EIAs generated intense political contests which widened to other issues, but the outcomes were much the same; the licences were renewed and increased.

Tasmania[17]

The EIA process, intended to evaluate single development proposals, was not well-suited to projects which had been operating for fifteen years. However, the EIA for the Tasmanian exports was even more unusual: one EIA was used to cover the three mills; the State's Forestry Commission, under instruction from the Tasmanian government, became a proponent jointly with the companies; and the proponents prepared their own EIS instead of engaging consultants to do so. In a further confusion, only a 'supplement' to the initial EIS, rather than a final statement, was ever prepared.[18]

The scope of the EIA was huge because it had to cover the three-quarters of Tasmania's state forests which were in the woodchip concessions, and the linkages with the sawmilling industry through the integrated regime. Hence the EIS, although lengthy, remained very general for most issues; it could not deal with impacts such as sedimentation in particular rivers or effects on particular sorts of wildlife habitat, for example. The proponents argued that woodchip export licences for 2.8 million tonnes a year should be continued, or increased if more wood could be found. They hoped to build a pulp mill to process some of the wood before export; they noted how forest practices had been improved, and promised to do better still, and to pay more royalties. Their sting was the threat of job losses if their plans were thwarted, a threat whose severity was exaggerated by a misleading study, yet which had enough substance to be politically powerful.[19] Notably, the EIS ignored the trade and price issues, and offered no feasible alternatives to continuing and enlarging the business.

The Forest Action Network (FAN) of national and Tasmanian environmental groups prepared a package of alternative proposals. They specified some forests with special wilderness or ecological values which should be protected from logging; they called for better forest practices, particularly on private property; and they called for an industrial development plan which would enable employment to be maintained, and made numerous suggestions for such a plan involving a higher degree of processing within Tasmania. FAN urged people to make submissions to the EIA, and provided pamphlets and a 'response kit' of information on how to do so. The response was substantial: almost two-thirds of the 460 submissions made were based on the kit.

The submissions typically covered several topics (Table 7.3), although some addressed local or very specific impacts, such as the effects of logging on freshwater molluscs, for example. The scope of the EIA was so large, and the number of submissions so great, that local or very specific matters received scant recognition in the

Table 7.3 Incidence and main character of topics mentioned in submissions to Tasmanian woodchip EIA

Topic	*Main character*	*Incidence (%)*
Alternatives proposed	FAN package	17.6
Economic aspects	Employment	15.4
Forest management	Sustainable yield, clear-felling	14.6
Wilderness	Destruction by forestry	12.7
Effects of harvesting	Fauna, water, soil, flora, steep country, archeology	11.5
Reserves	Various	9.4
Procedural	EIA	8.4
National Estate	Various	6.9
Other		3.5
Total		100.0

Source: Calculated from Tasmanian Woodchip Export Study Group (1985b), pp. 233–6.

Commonwealth's assessment, let alone by the woodchip companies. They, with the Forestry Commission seemingly in tow, issued a supplement to their EIS which clarified some of the statements and rebutted, rather than made adjustments for, criticism raised in the submissions. They noted FAN's suggestions for adding value in processing, but argued that any such improvements should be directed to economic growth rather than to reserving forests from logging. The supplement offered nothing to the environmental groups. The contest moved to the federal arena, where the EIS had to be 'assessed' and a political decision made. The decision was not so much about renewing the licences – that was never really in doubt – but about which areas of forest the State would reserve from logging as a condition of the Commonwealth issuing the licences. The outcome was to increase the export quotas slightly, improve practices, and reserve rainforest and some National Estate areas from logging under a Memorandum of Understanding (MOU) between the two governments. In essence, most of the forests were to continue to be refashioned by woodchipping.

South-east New South Wales

Although the EIA for Harris–Daishowa's woodchip mill at Eden had to deal with only one mill and was more competently prepared by consultants, the outcomes were much the same: no 'prudent and feasible' alternatives were identified, there was an intense political contest over boundaries, logging was stopped on some National Estate areas, the licence was renewed and increased, and most of the forests continued to be refashioned by woodchipping.[20] Rather than join the company as a proponent, the Forestry Commission of New South Wales tried to stay at arm's length, believing that its management plan and logging prescriptions defined its responsibilities and intents clearly enough. Behind this was the commission's right to manage 'its' forests independently of both the Commonwealth and the NSW Department of Environment and Planning. Yet, the major environmental impacts were in the forests; those around the mill itself were trivial by comparison. In consequence, the proponents could not give any undertakings about how the forests would be managed, and hence what the impacts would be. It would not do. The Commonwealth pressed the New South Wales government; the director of the Department of Environment and Planning 'made known his requirements'; and some months later the commission issued a supplement to the draft EIS.[21]

The South-East Forests Alliance (SEFA) encouraged people to make submissions; there were 302 from individuals and organisations covering the entire range of interests. The environmental submissions concentrated on having areas of high ecological or wilderness value reserved in national parks, or registered as part of the National Estate. In particular, they sought to add two large areas, at Tantawanglo and Coolangubra, and some smaller ones to the existing parks and reserves, so that the region could have a larger and more representative system of reserves. Various alternatives to woodchipping were devised to offset consequent job losses.

The final EIS was duly lodged at the end of 1986.[22] It concluded that the industry should continue, that compliance with logging prescriptions should be improved, and that more ecological research should be undertaken. From there, the contest rapidly escalated. The Forestry Commission pushed roads into the National Estate areas and started to log them, only to be met with massive protests in the forests organised by SEFA. Mobs of people stood on the roads to stop the log trucks, a few erected platforms on which they lived high in the trees to prevent them being felled, and rallies were held in Canberra and other major cities. The New South Wales government reacted by closing the roading and logging sites to the public and sending a contingent from its Tactical Response Group to reinforce its local police. Their 'Operation Redgum' lasted for four months, during which more than 1000 people were arrested in a generally non-violent manner.[23] The timber workers and loggers, sick of the disruptions and threats to their livelihoods, protested too by holding a rally in front of Parliament House in Canberra and by blockading environmentalists' access to the disputed areas.[24]

The contest raged politically within the New South Wales government as its National Parks and Wildlife service, hurried by its Environment Minister, recommended that Coolungubra be declared as a wilderness, only to have the Resources Minister persuade the government to reject it.[25] It raged within federal government as the Environment Minister, Senator Graham Richardson, recommended that there be no logging in National Estate forests. And it raged between the two governments as the federal government placed a temporary ban on logging National Estate areas which the Forestry Commission promptly broke.[26] Eventually they agreed to suspend logging in 90 per cent of the contentious areas while a joint study was conducted. Nevertheless, the Forestry Commission pushed ahead with its construction of the Wog Way Road, which resulted in further protests, tree-sitting and arrests. Finally, the licence was renewed and its quantity increased with some wood from East Gippsland. Logging was stopped in part, but not all, of the Tantawanglo and Coolangubra areas, and the New South Wales government promised to place them into national parks.

The Commonwealth and New South Wales Ministers prepared a *South-East Forests Agreement* in 1990 to cement these decisions and establish a regional advisory committee to advise how $10 million allocated by the Commonwealth for research, industrial restructuring and adding value to the wood should be spent. The agreement was ratified by the federal cabinet in 1993, but stalled in New South Wales in the face of an independent member's Bill which sought to transfer a further 96 000 hectares to national parks. Doubtless, the refashioning of these forests will continue to be contested in these and many other ways while clear-felling on such a large scale continues.

Patterning the forests and ameliorating the impacts

The environmental outcry against the impacts of woodchipping induced the forest services to change the pattern in which the forests were felled, and prescribe various 'codes of forest practice' to reduce the environmental impacts of logging.

Changes to the pattern of felling involved reducing the size of the logging coupes and distributing them more widely across the landscape. The most marked change was in the south-east forests of New South Wales which supplied the Eden woodchip mill. At the start, logging coupes there were as large as 700 hectares and were set out in a continuous network. Their size was reduced first to about 200 hectares and then typically to about 30 to 50 hectares set out in a patchwork pattern, so that areas being cut were always separated by others which would not be cut until twenty or so years later. Within each, patches of advanced regeneration and strips along each creek were left uncut, and a few large 'habitat' trees were retained on each hectare to provide the hollows in which birds, possums, bats and other creatures could have their being.

Soil erosion and the consequent deterioration of water quality was addressed by leaving 'filter' strips of uncut vegetation to trap the sediment running off the coupes and roads, and by leaving patches of the forest uncut where the slopes were steep, especially if the soils were particularly erodable. After logging was finished, the contractors had to 'bar' their tracks to prevent erosion and rip the ground on log landings to counter its compaction. The Soil Conservation Service gradually increased its influence, made its *Soil Erosion Mitigation Guidelines* more demanding, and coupled them with training for foresters and loggers. The detailed prescriptions for all these 'forest practices' were strengthened from time to time and enforced more thoroughly.

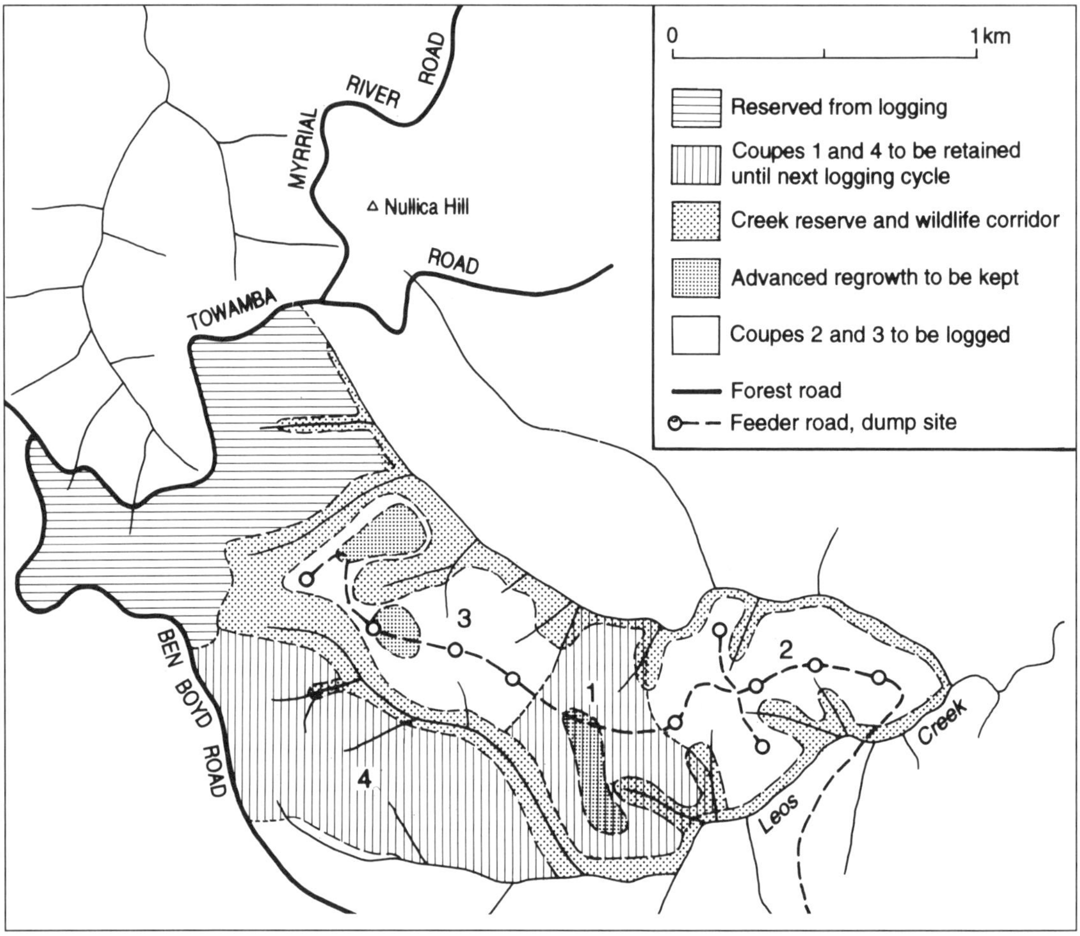

Figure 7.1
Compartment harvesting plan
Source: Australia, ESD Working Group on Forest Use (1991).

Victoria had developed detailed prescriptions for controlling logging in declared water supply catchments since 1950. These were applied more widely and progressively improved until, in 1989, a *Code of Forest Practice* was ratified. In 1985 the Tasmanian government, needing to bolster its case for continuing the woodchip industry, introduced a system based on detailed planning and industry self-regulation, but set far less stringent standards than those in Victoria.[27] In the other States, logging standards were prescribed by variously named 'codes' and 'guidelines'. Although it took time for the loggers and supervisors to adapt, and planning, compliance and enforcement were never perfect, many of the immediate and local impacts of logging were gradually reduced.

But the long-term and broad-scale ecological impacts of extensive clear-felling across the forest landscape were inescapable.[28] Most obviously, the woodchip concessions were being transformed to forests of regrowth dotted with retained habitat trees, and inscribed with strips of retained vegetation and patches too steep for logging. By the 1990s this process was about half completed; when it is completed in a further twenty years or so, their forests will consist mainly of even-aged stands of trees under about forty years, except for the retained trees, strips and patches. Ecologically they will be a very different from the forests they replace; their great blocks with old trees – some at Eden 400 years old and many 200 years old – will be gone, and the habitat trees and strips retained will inevitably diminish with age, fire, tempest, and the intrusion of weeds and vermin.[29] As ecological research advanced, it was found that the development of the large hollows needed by some species for nesting only occurred in most species of trees after they were more than 120 years old and it continued in some for centuries. Moreover, habitat trees would not be enough for some species, survival if the understorey habitat and food species were lost.

Forest clearing for pine plantations

The rapid spread of pine plantations from the mid-1960s affected about the same area each year – 33 000 hectares during the 1970s – as woodchipping, and created many scenes as hideous. It drew savage criticism from the Routleys.[30] The heart of the issue lay in the fact that the forest services were clearing areas of native forest in order to plant pines. This was sensible from their wood production perspective, because the pines could grow faster than the native forests and the services already had the land, but it took no account of environmental values. In New South Wales, the Forestry Commission was pushing its plantations into montane forests which were also being affected by woodchipping. Victoria and Queensland were also pushing plantations into native forest regions. For example, the sandy 'wallum' region in south-east Queensland which had negligible commercial value for hardwood timber was being cleared for both agriculture and pines. In the 1970s, the need to preserve some of its remaining flora and fauna started to be asserted and the pine program was opposed.

The impacts of clearing native forests for pine plantations were more severe than woodchipping, and it was feared at the time that successive rotations of pines would ruin the soil.[31] In an early move, a House of Representatives committee included the impact of plantations during a

Eucalypt forest cleared for a pine plantation, Gippsland, Victoria, 1973. (John Dargavel)

wider inquiry on wildlife.[32] In a dramatic phrase of the time, the pines were depicted as 'biological deserts', although in fact many native species of animals and birds survived in whatever odd patches of native vegetation had been left along creeks.[33]

Decision

Many of the environmental impacts could be reduced by improvements in plantation practices, without changing the substance of the program, and these the forest services and some of the industrial plantation owners adopted fairly readily. Strips of native vegetation were left along roadsides to soften or even hide the visual impact, gullies and creeks were left as habitat, various patches preserved, and 'corridors' or 'buffer strips' of vegetation kept so that wildlife could move across the landscape. While this ameliorated the immediate impacts, the long-term survival of narrow strips of native vegetation through pine plantations is doubtful.

The ACF and many other environmental organisations proposed that land which was uneconomic for farming should be bought for pine planting rather than clear native forests.[34] Gradually the climate of opinion changed. In New South Wales, for example, the Australian Labor Party while trying to gain government in 1976 stated that it would 'prevent further clearing of native forests for pine plantations'.[35] In power, the money to buy land was not so easily found, nor was the Forestry Commission so easily turned around. Nevertheless, by the mid-1980s all the forest services had stopped clearing extensive areas of their native forests and were buying land for their plantations.

Partly in response to the rising cost of land, plantation silviculture became far more intense from the 1970s as the techniques of soil cultivation, fertilisation and weed control were progressively advanced, and tree breeding improved growth rates and tree form.[36] The great era of complaint against pines passed, with little attention being paid to the environmental effects of such intense silviculture.

National parks and state reserves

The parks associations and other environmental organisations in all States campaigned with considerable success during the 1970s and 1980s for large areas of forest to be reserved as national parks, while many smaller areas were also set aside for recreation and conservation in other types of reserve.[37] The area of forests in parks and reserves in Australia increased four-fold, from 1.8 to 6.2 million hectares, mostly from unallocated Crown land, while the area of state forests was also increased slightly (Figure 3.1).[38]

National parks

The long-established practice of reserving attractive single areas as national parks on land generally unwanted for agriculture, mining or timber changed towards an ideal of reserving areas which together could provide a system representative of all of Australia's ecosystems.[39] Many of the claims made for new parks were for large areas, often rich in timber resources. Indeed, the remaining unlogged areas in prime forest types – often poorly represented in the older parks – were claimed urgently to prevent impending roading or logging. The area of parks and reserves was also increased by various reviews and by additions to existing parks, sometimes by purchasing private land through bodies such as the National Parks and Wildlife Foundation in New South Wales (Figure 7.2). To manage the larger areas, all the States and Territories set up National Parks and Wildlife services between 1967 and 1978.[40] The Commonwealth set up its service in 1975, largely as a policy and funding body. The new parks services, imbued with the cause of environmentalism, had to struggle for real as well as bureaucratic territory with the established forest services.

In 1977 the Premier of Victoria announced that 5 per cent of the area of the State would be reserved for conservation.[41] As Victoria's Land Conservation Council proceeded with its work round the State, more parks were declared, so that the area of forest land included

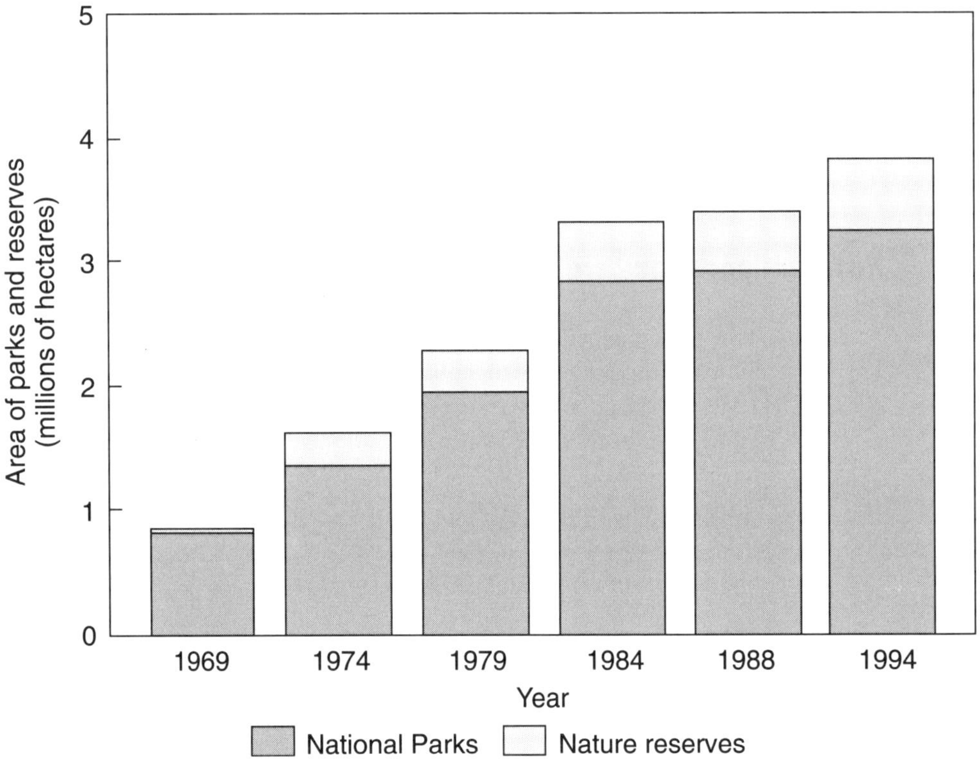

Figure 7.2 National parks and nature reserves, NSW, 1961–94
Source: ABS, *Official Year Book of New South Wales*, Sydney, various years.

rose from 128 000 hectares in 1971 to more than 1.6 million hectares by 1991.

State reserves and management priorities

The States reserved many smaller areas of state forests for flora and fauna reserves, historic sites, recreation, scientific and many other purposes. New South Wales, for example, gazetted 153 flora reserves, ranging from 12 to over 700 hectares, to be managed by its Forestry Commission.[42] Areas were also set aside from logging, or treated specially, by agency policy or planning. The Forestry Commission in New South Wales, again for example, delineated sixty-two 'forest preserves', averaging 250 hectares each, as scientific reference sites, some of which, after assessment, were gazetted as flora reserves. It also developed a 'Preferred Management Purpose' (PMP) classification in its forest planning system in 1984, which delineated parts or patches of the forest with particular values that had to be managed in special ways. The PMP system was used for the Eden area's plan in 1982. Victoria adopted a very detailed planning system which delineated patches of the forest, often as small as 5 to 10 hectares, for special management. Similar changes were made in the other States.

World Heritage and National Estate

World Heritage

Ten Australian sites were listed as being of world importance between 1981 and 1992, of which four containing significant areas of forest were the subject of great environmental contests (Table 7.4).[43]

Tasmania's Lemonthyme and Southern Forests[44]

The difficulties of delineating World Heritage areas were exposed most sharply in south-west Tasmania, where the preservation of great undeveloped areas had been advocated from the 1960s. Although the Tasmanian government had dammed Lake Pedder, it had also created some national parks and proclaimed a large Conservation Area which was listed on the Register of the National Estate in 1980.[45] Its Western Tasmania Wilderness National Parks of about 700 000 hectares had been registered as World Heritage in 1982, but there were further areas in the Lemonthyme Forest and others to its south, which had already been included in the Register of the National Estate, which warranted nomination.[46] They contained the best unlogged mature ash and

Table 7.4 World Heritage properties with significant areas of forest

Date listed	*Site*	*State*	*Area (hectares)*
1982 and 1989	Tasmanian Wilderness Lemonthyme and Southern Forest areas added	Tas.	1 374 000
1986	Australian East Coast Temperate and Subtropical Rainforest Parks	NSW	203 564
1988	Wet Tropics of Queensland	Qld	890 000
1992	Fraser Island	Qld	166 283

Source: ERIN (1994), *Managed Area Reports*.

messmate stands left in the State – indeed, in Australia – and hence were highly prized for both their ecological and commercial values.

Rather than nominate the region immediately, the Commonwealth set up an inquiry, presided over by a retired judge, Michael Helsham, to determine how much should be nominated and what 'environmentally and economically prudent and feasible alternative' resources were available for the timber industry. The Helsham inquiry commissioned consultants to study the 284 000 hectares, heard expert witnesses and the cases of both environmental and industry groups, and issued an interim report that a small area of 226 hectares could be logged immediately. But in their final report, the three commissioners could not agree on the main issue; two found that only 34 850 hectares satisfied the criteria for World Heritage listing, one found that the entire area plus some adjoining land did.[47] Most of the scientific experts felt that the formal processes of the judicial inquiry had proved unsuited to heritage assessment, while the Commonwealth was miffed that the whole issue was back in the political arena. After intense lobbying, several rounds of negotiations with the State, a change of State government and the negotiation of the Salamanca Agreement there, the Lemonthyme and Southern Forests area and uncontentious areas to their west were nominated and accepted on the World Heritage list in 1989.[48]

Fraser Island[49]

The suitability of Queensland's Fraser Island, and surrounding areas of the Great Sandy Region, for nomination as a World Heritage area was also the subject of an inquiry, but by the State. The island, about 120 kilometres long, is the largest one made only of sand in the world; it has unusual freshwater lakes and tall rainforests which had been logged selectively since the 1870s. Although the mining of mineral sands there had been stopped in 1976 after a long contest and a Commonwealth inquiry, the Commonwealth was reluctant to enter into major dispute with another State government as it had in Tasmania. Hence, it was not until 1989 when the Goss Labor government had defeated the long-entrenched National Party that environmental and heritage matters could advance. Rather than implement its party's policy of declaring the island a national park immediately, the cautious new government called for an inquiry to be conducted by a single commissioner, Tony Fitzgerald.

It was less formal than the Helsham inquiry but had to consider some matters beyond the impacts of tourism and logging on the island itself. Eventually, it reported that the region met the criteria for World Heritage listing and recommended that logging be stopped.

Queensland asked the Commonwealth to nominate the area, logging was stopped, and the timber industry compensated. The island was made into a national park and, in 1992, was listed as a World Heritage area.

National Estate

Fraser Island was the first place to be registered as part of the National Estate in 1977, but by 1993 over 10 000 more sites had been added of which 16 per cent were natural ones (Figure 7.3).[50] More than half the natural places were national parks, reserves, wilderness, and habitat for endangered species, categories in which the native forests were prominent (Figure 7.4).[51] Overall, one-quarter of the nation's forests were listed, the proportion varying from State to State (Table 7.5).

Many of the listed forests were outside national parks and conservation reserves, so that their natural values were contingent on how the States recognised and managed them. Listing only obliged

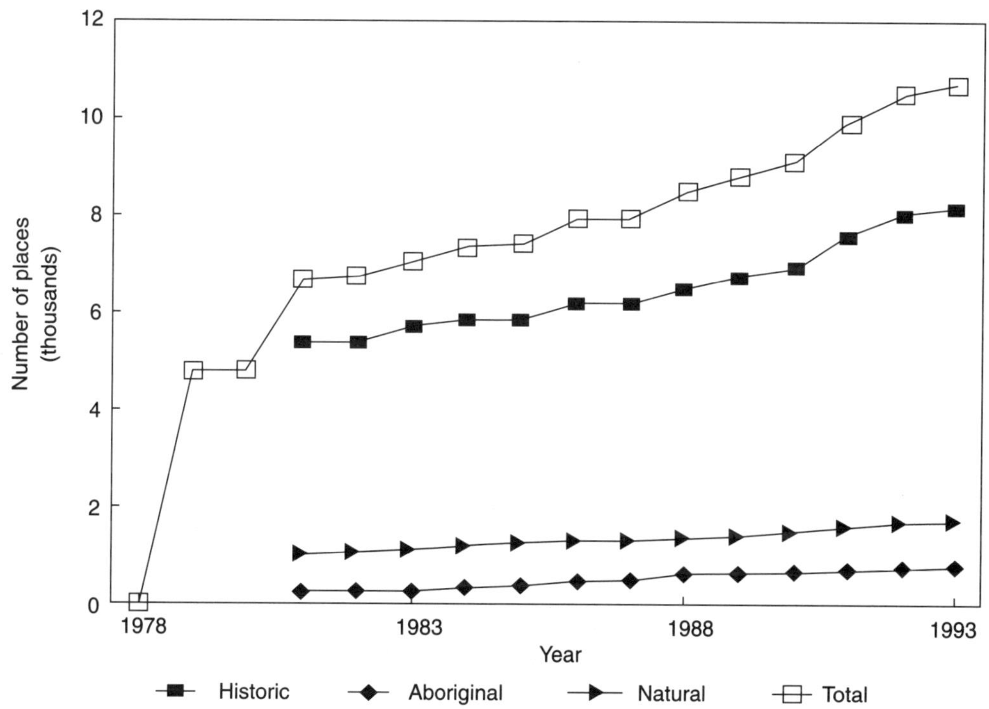

Figure 7.3 Number of places in the Register of the National Estate by major group, 1978–93
Source: AHC, *Annual reports.*

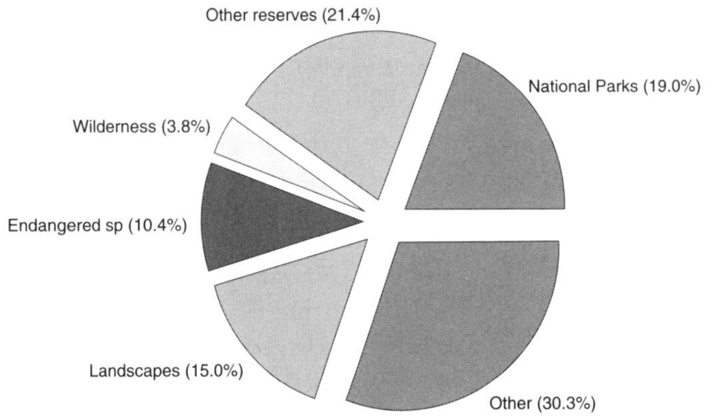

Figure 7.4 Proportion of natural places in the Register of the National Estate by category, 1993
Note: 'Other' includes arid, alpine, marine, geological, wetland and miscellaneous categories.
Source: AHC, *Annual report*, 1992–93.

Commonwealth Ministers to consider heritage values in their decisions, such as granting woodchip export licences. However, it effectively induced State forest services to give special attention to heritage values in their management. Listing proved a powerful means for environmental groups to bring their values to public attention; certainly, the forest industry and some services protested that listing was becoming a *de facto* method of land-use planning due to the 'intense political pressure' which they said listing 'provoked'.[52] A more systematic and harmonious process was needed.

Western Australia's Department of Conservation and Land Management and the Heritage Commission cooperated in assessing national estate values in the karri belt in the south-west of the State in 1990.[53] Working systematically on a regional basis and using well-defined, consistent criteria, they were able to avoid the previous *ad hoc* approaches and provide the basis for agreement between the two governments. This regional assessment model was adopted by the Victorian government for a thorough study of the East Gippsland and Central Highlands regions and is being extended to other States.[54]

Wilderness

Wilderness was a more exacting claim than those made for natural heritage or national parks, yet it was one whose mix of spiritual, recreational and ecological aspects eluded any final definition.[55] It had no meaning for Aboriginal people to whom the land was a cared-for home, but to non-Aboriginal people wilderness offered the experience of solitude, the challenge of self-reliant adventuring, and ecologically it provided areas whose natural patterns and processes were not disturbed by humans. Only large, remote, untrammelled tracts of land would do; but just how large, remote, or untrammelled was a matter for judgement.

In 1974 the Colo Committee was formed to press for the protection of 500 000 hectares in the Blue Mountains; in 1975 the ACF proposed that a 'nation-wide system of wilderness reserves' should be created; in 1976 Victoria's LCC delineated wilderness areas in the Mallee (114 000 hectares) and East Gippsland (46 000 hectares); in 1977 the first of a series of national wilderness conferences was held in the Academy of Science; and in 1979 the Colo area was proclaimed as the Wollemi National Park with wilderness at its centre.[56] But where else was there wilderness? The first systematic survey was made in New South Wales and the ACT in 1976, with funds from the Heritage Commission.[57] It confined the elusive nature of wilderness in a workable definition, set standards for mapping, and delineated twenty areas between 20 000 and 235 000 hectares.[58] Of these, three in the north coast region – Barrington, Mann and Washpool – contained valuable rainforests, described in the next section.

The Franklin dispute followed by the Helsham inquiry put wilderness on the national political agenda and made the quest to find and map it seem an urgent one. A National Wilderness Inventory was set up in the Heritage Commission in 1986, but it had to develop ways to identify wilderness and then obtain data, mostly from State agencies. What emerged was a fluid idea of wilderness which would change 'in response to the patterns of community demand for remote and natural

Table 7.5 National Estate forest places and areas, 1990

State	*Number of places with forest*	*Proportion of State's forest listed in the National Estate (%)*	*Proportion of National Estate not in conservation reserves (%)*
NSW	208	14	17
Qld	175	42	70
SA	95	100	33
Tas.	67	25	71
Vic.	94	25	16
WA	53	26	10

Source: Adapted from RAC (1992), *Forest and Timber Inquiry Final Report*, vol. 2A, p. 16.

land and its supply', and four indicators (remoteness from settlement and access, and apparent and biophysical naturalness) which could be mapped and combined to show various qualities of wilderness (Figure 7.5).[59] Gradually, wilderness gained greater official recognition. New South Wales, for example, legislated in 1987 to require its Parks and Wildlife Service to assess the potential wilderness areas of the State. By 1993 wilderness had been mapped and set aside under State legislation, or agreements with the Commonwealth, in Tasmania, Victoria, South Australia, Cape York and the wet tropics of Queensland.[60]

Rainforests[61]

The claim to preserve rainforests as rich, rare, lovely and threatened had particular force: they are biologically profuse, containing 20 to 30 per cent of all the plant species in New South Wales and Queensland even though they only cover less than 1 per cent of the land; there are only 2.3 million hectares left; only small areas, such as Lamington National Park and various flora reserves, had been preserved; and the small but conspicuous areas in private hands were subjected to heedless development. Except for the steepest slopes, most of the rainforests either had been, or were intended to be, logged. This had patterned them in stages, each reaching further and taking more species of trees.[62] By the 1970s the forest services exercised considerable control over the selection and felling of the trees, but it was too late: the mill licences were for double the quantity which could be sustained. Logging had a poor history, and even the rarest and most valuable species were being sold for a pittance; forestry was more promise than proved performance.[63] Importantly, there were still stands of 'virgin' rainforest in remote sites which were to be opened up to maintain the declining industry.

The ACF urged that commodity producing activities which 'adversely affect' rainforests should be phased out so that 'all areas of prime undisturbed rainforest' could be reserved in national parks.[64] It conducted a national study and convened a major conference in 1983.[65] With mounting public concern, the Commonwealth too convened a conference and in 1984 set up a Working Group to see what the national policy should be.[66] The claim for rainforest preservation was refined in such studies. They contrasted the biogeography of the rainforests with the rest of the dry continent, catalogued the abundance of plants and creatures, celebrated their natural values, and simply held logging or any other human disturbance to be self-evidently

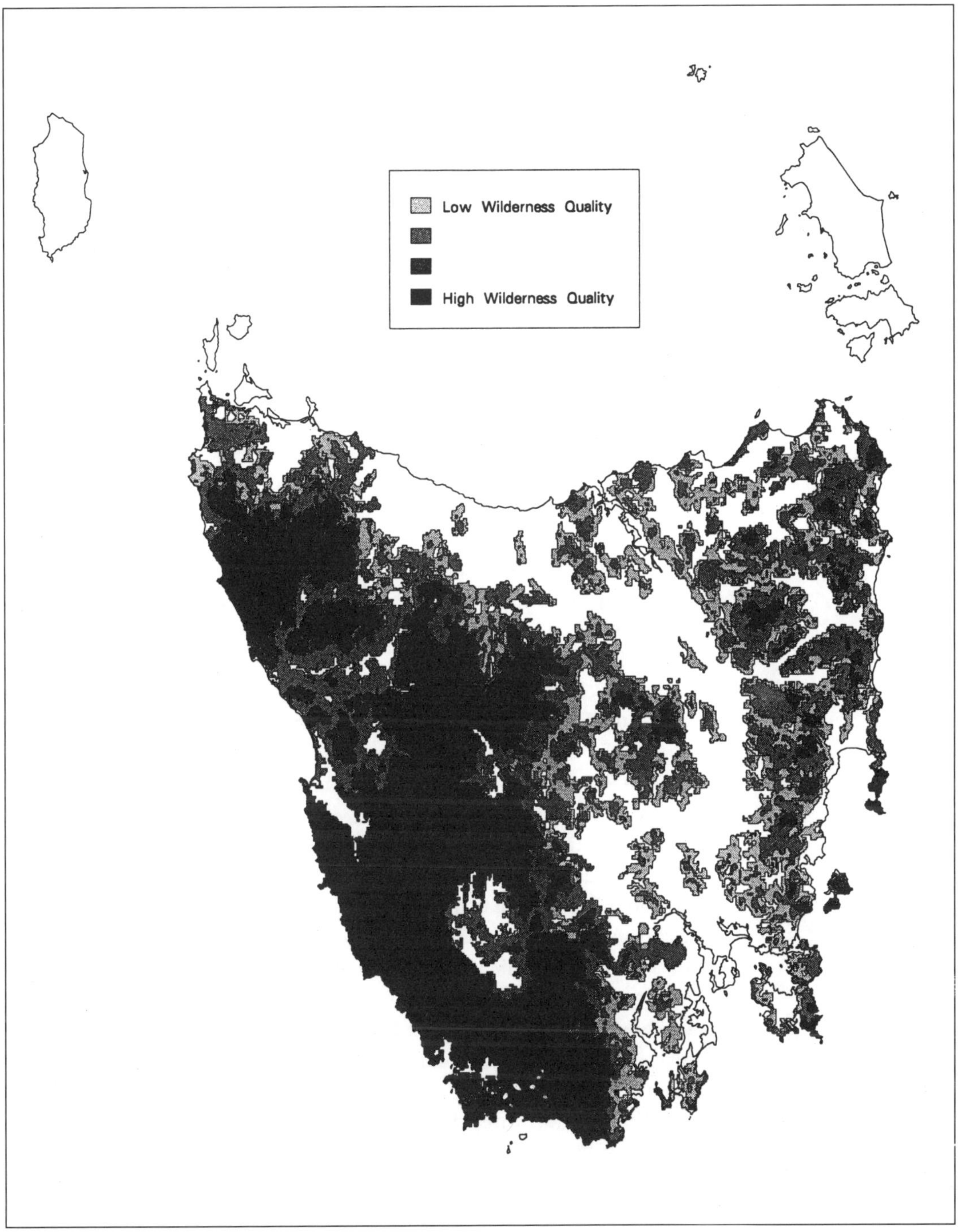

Figure 7.5 An example of wilderness mapping
Source: M. Maslen, AHC.

bad. The ensuing contests in New South Wales and then in Queensland were echoed in Tasmania and Victoria.

New South Wales

New South Wales has only about 250 000 hectares of sub-tropical and warm temperate rainforests left. They lie in six separate groups, south from the Queensland border to the Hunter River. Three in remote and rugged country contain the wilderness areas mentioned earlier. All are intricately intermingled with strands of wet eucalypt forest, itself a rich habitat sharing some understorey species with the rainforest. Small patches of rainforest vegetation can also be found in wet gullies south into Victoria.

The contests started in 1973, when a local environment group proposed that some 33 000 hectares of state forest in the Border Ranges should be turned into a national park which would complement Queensland's adjacent Lamington Park.[67] The proposal was taken up by the Colong Foundation, the Royal Society of Queensland, ANZAAS in New South Wales, and then nationally by the ACF. It was opposed by the Forestry Commission which, faced with severe declines in supplies from its more accessible forests, was intent on roading and logging the area. Official inspections and inquiries were conducted for years, during which time the public and political support for environmental claims steadily rose. Even though the government changed to a more sympathetic Labor Party, the matter seemed deadlocked in a dispute between the State Pollution Control Authority and the Forestry Commission.

The situation was suddenly changed – not just in northern New South Wales, but for forests throughout Australia – by a direct confrontation which erupted over quite a small basin of some 700 hectares at the head of Terania Creek, 25 kilometres south-east of the Border Ranges. Ironically, the basin was only partially rainforest and the Forestry Commission had planned to cut only 2000 brush box and blackbutt trees on its margins. But to do so it intended to build a road, partly through the rainforest, to facilitate the logging, which would obviously break the natural peace of the basin as a whole. The region, near Nimbin, was one which had attracted many 'new settlers' by its warmth and natural beauty. Their small Channon Residents Group had argued to little avail for years that the basin should be made into a nature reserve. When the Forestry Commission started to build the road in 1979, the group was joined by hundreds of protesters who lay in the path of the bulldozers, jumped on trucks, sang songs, and some spiked trees.[68] It was the first forest blockade. The commission huffed,

Natasha on tree platform 25 metres above the ground: 'You do some very intense thinking in this place'. Protest against felling in Coolungubra Forest, Eden woodchip concession area, 1989. (Senka Stemberger)

the industry puffed, the police came, arrests were made, and it all made splendid news for the television cameras. The government set up yet another inquiry which eventually recommended in 1981 that Terania Creek could be logged.

Meanwhile, contests extended to other groups of rainforest in the Hastings, Black Scrub and Washpool areas, of which the Washpool contained the largest area of untrammelled wilderness. The contests entered the legal and bureaucratic arenas when the commission's operations were brought under the provisions of the State's new *Environmental Planning and Assessment Act 1979* which required it to prepare EIS for new developments. And the contests drew in more players, ranging from the National Trust and professional bodies to the Total Environment Centre and the Nightcap Action Group. They published, lobbied, and protested to good effect, causing popular pressure to 'save the rainforests' to mount. Finally, in 1982, the Cabinet declared that the six main areas of rainforest would be put into national parks. Over the next four years, part of Washpool was declared a wilderness area within the park, and the State government successfully pressed for all six areas to be nominated for the World Heritage list.[69]

Queensland[70]

Although Queensland has slightly more than half of Australia's rainforests, contests over their use occurred later than in New South Wales and their eventual listing as World Heritage was strongly opposed, rather than initiated, by the State government. Many of Queensland's rainforests had been cleared for agriculture, especially for growing sugar on the best tropical lowlands. Clearing continued in areas such as the Atherton Tablelands into the 1950s and for speculative residential developments into the 1990s.[71] As in New South Wales, what rainforests remained were largely too remote or too steep for agriculture. The main wet tropics area of 790 000 hectares lay between Townsville and Cooktown; further north on Cape York Peninsula were 260 000 hectares, mainly in three separate groups; while a further 200 000 hectares were spread out south to Lamington on the border in about fifty separate areas.[72] One-quarter of the rainforests had been placed in national parks, and slightly over half were in state forest reserves.

The principal contest was over the wet tropics.[73] It had a quarter of Australia's plant genera, the highest concentration of primitive plant families in the world, and 60 per cent of Australia's species of bats, for example. The valley of the Daintree River at its northern end was the centre of the largest tropical rainforest wilderness in Australia, and the association, near Cape Tribulation, of fringing coral reefs with rainforest growing to the water's edge was unique in Australia, and hardly found elsewhere.[74] It was beautiful, a jewel of the Australian environment.

The rainforests also supplied logs to thirty small sawmills which cut timber for the local market and sent cabinet timbers south. They had been allowed to cut more than 200 000 cubic metres a year from state forests, but by the 1980s resources had dwindled so far that they were reduced to less than one-third of that. Logging was planned to take only the largest trees and to return every forty years or so when the rest had grown. Although the multi-layered structure of these forests recovered, their size-age structure was changed. The impacts were subject to hot discussion: the foresters claimed they could manage sustainably and log carefully, but environmentalists pointed to where logging had smashed surrounding trees or caused erosion; foresters saw that such things healed naturally, but environmentalists pointed out that the ecology of the rainforests was so complex that the long-term impacts of logging were unknown, even though the foresters pointed out that no species had been lost. The environmentalists' most urgent goal was to stop roads being pushed into ever more remote areas with 'virgin' stands, such as the Windsor Tablelands.

The Heritage Commission put 350 000 hectares of the Greater Daintree region on the Register of the National Estate in 1980 and the Queensland government declared a national park at Cape Tribulation the next year, but logging proceeded and had broken into half the Windsor wilderness by 1983.[75] However, it was not logging but the crass bulldozing of an unsurveyed four-wheel-drive tourist road through the national park and the Daintree wilderness area which suddenly changed the situation. Blockades to stop the road in 1983 and 1984 were roughly over-run by the Queensland police, all of which again made the national news.[76] Demonstrations, protests, lobbying and careful detailing of the scientific information by the Rainforest Conservation Society of Queensland continued until, as part of the 1987 election, the Commonwealth nominated the main area for World Heritage listing. This empowered it to stop further development, although it was too late to stop the road.

Tasmania and Victoria

The claims to preserve rainforests were extended to the cool, temperate rainforests of Tasmania, which, with a few small areas in Victoria, make up about one-third of Australia's rainforests. They have comparatively few species of either animals or trees, of which the most noteworthy are the slow-growing huon, celery-top and King Billy pines endemic to Tasmania, and blackwood and myrtle beech which provide furniture timbers. Many of the stands of these rainforest trees have occasional large eucalypts 'emergent' above their canopy due, it is thought, to their dynamic patterning over long periods of time. If burnt, they regenerate to the quick-growing, taller eucalypts; if not, the slower, longer-lived rainforest species form a dense canopy which prevents the eucalypts regenerating when at last they die. In such stands, any distinction between 'rainforest' and 'eucalypt forest' was ecologically meaningless, though it was potent politically; 'rainforest' had become a public icon to be saved.

Both States produced special rainforest conservation policies. Tasmania's prohibited logging and burning in rainforest, but allowed patches up to 10 hectares or so to be felled and regenerated with eucalypts if they were in coupes of mostly eucalypt forest. Policies for huon and King Billy pines permitted some of their special timbers to be cut for the furniture and craft industries. Blackwood was more plentiful and was covered by a special plan for its cutting, regeneration and plantations. Victoria, with only 13 000 hectares in small scattered patches, of both cool and warm temperate types, not only banned logging within them, but sheltered them with surrounding buffer strips where necessary.[77]

Endangered species and biological diversity

Rare and endangered species

The fundamental claim that all species have an inherent right to exist was not to be gainsaid, especially in its most urgent aspect of saving those in danger of extinction. In 1987 the Commonwealth called a national conference on threatened species and set up an Endangered Species Unit, while the States strengthened their flora and fauna legislation.[78] The national catalogue of endangered or vulnerable species listed 2500 plants, 50 birds, 46 mammals, and sundry frogs, reptiles and fish.[79] The greatest immediate threats to those in the forests came from clearing or grazing, but extensive clear-felling posed long-term threats too. As the scientific evidence accumulated, the basis for much conservation had to shift from reserving strips and trees as 'islands' of appropriate habitat to analysing how whole forest landscapes should be managed to maintain viable populations of animals such as the koala, Leadbeater's possum and the long-footed potoroo.

The much-loved koala was at risk in New South Wales, South Australia and Queensland, while in Victoria it had been saved from extinction only by reintroducing animals from some island populations.[80] A national survey in 1986 and 1987 clarified that its plight, due to past land clearing, the fur trade and disease, was being made worse along the north coast of New South Wales by land developments which cleared the fertile coastal sites on which it did best. Its survival would have to depend largely on sparse populations scattered through forests, such as those in the south-east being logged for woodchips. One area there was temporarily set aside until a management plan for the region could be prepared, but this stalled in squabbles between the forest and wildlife services.

Leadbeater's possum was thought to be extinct until it was rediscovered in the prime mountain ash forests of central Victoria in 1961. Its tiny population of some 4000 animals is very vulnerable to losing the old hollow trees in which it builds nests to raise its young. Its domain, ravaged by the 1939 fires, still contained plenty of such trees in the 1990s, but they must fall in time so that its survival depends on their being replaced by other ageing trees. Until the 1980s it was thought that the areas reserved from felling in Melbourne's water catchments and the odd trees left through the commercial forests on steep slopes or stream edges would do, but research found that this was not so. Additional blocks of trees would have to be left grow beyond the 80–120 years for which felling was planned, to 150–200 years, with some perhaps left even longer. Such blocks would have to be arrayed in patterns which will permit the little animals to colonise them from adjacent areas when the time is ripe and ensure that they could survive great bushfires again. Such patterning is far different from that which is needed for timber management. After some years of discussion with wildlife biologists, the timber industry and forest managers, the pattern of cutting was changed, but the radical step of greatly lengthening the age of clear-felling was not taken as it would have considerable effects on the timber industry.[81]

The shy, little long-footed potoroo is one of the rarest mammals in Australia, with a small, scattered population in East Gippsland and south-east of New South Wales. Patient research found that it came out at night to dig for fungi on the roots of trees and shrubs. It found them from ridge to gully at different times of the year, so that its conservation demanded retaining all the vegetation, not just streamside strips.[82] In Victoria some of its habitat was already in national parks, and elsewhere its habitat was reserved from logging wherever it was found. In south-east New South Wales it was so rare that when their hairs were discovered in traps in 1986, the industry thought that the environmentalists had planted them! However, the most intensive survey ever undertaken was carried out over 30 000 hectares using 13 000 tube traps to catch their hairs – the standard method. Only six positive, one probable and fourteen possible records of its presence were obtained.[83] A plan to protect them will be prepared.

Biodiversity

The preservation of endangered species was a politically potent claim when they were portrayed as appealingly as these three examples. But ecologically, it was a partial claim – what of the myriads of invertebrates, for example? – beyond which lay the greater claim to ensure the ecological functioning of the forest as a whole. It secured a serious scientific base and a place on national and State agendas under the banner of biological diversity – 'biodiversity'.[84] The scientific concerns were gathered under a new heading of 'conservation biology', while a 'new forestry' was advocated to deal with 'ecosystem management'. The possibilities for realising such ideas are discussed at the end of this book.

Continuing conservation

From the foresters' perspective of 1970, the new parks and heritage areas, strips and corridors, nesting trees and changed logging practices were substantial responses to the environmental claims. Yet from the environmentalists' perspective of 1994, the conservation of the ecological processes of the forests and the existence of wilderness and endangered species, such as those just cited, are continuing claims for which much remains to be done. Most seriously, much of the subtle ecological fashioning of the forest is still to be revealed, new creatures discovered even, and new claims made. From the industrialists' perspective, the environmentalists' claims conflicted with their own to the resources and fettered their investments with uncertainty, as discussed in the next chapter.

CHAPTER EIGHT

CONTESTING INDUSTRIES AND REGIONS

The virtue of the environmental claims gained hold on the public mind but was irrelevant to capital, except as an impediment to profit. Industrialists and the forest services first objected to many of the measures to ameliorate the environmental impacts of their logging, then strongly opposed the 'locking up' of timber resources in conservation reserves. In the mid-1980s they proposed a plan for industrial expansion to be led by the construction of 'world-scale' pulp mills designed primarily for export markets. When this foundered and even existing pulp mills saw their resources shrinking, they proposed a new legislative framework of 'resource security' to protect them. They also proposed to expand the area of plantations, as did environmental groups for different reasons. This chapter describes the contests over these proposals.

Organising defence and offence

Industry defended what it saw as its historic 'rights' in the native forests, contesting the mounting environmental claims in its traditional manner by lobbying State politicians. Some attempts to influence public opinion were made in 1973 when Dr Max Jacobs, highly-respected former director-general of the Forestry and Timber Bureau, was engaged to explain the benefits of forestry to Ministers and media under the slogan of 'a managed forest is a happy place'. Shortly after, the first formal defence organisation, the Forest Industries Resource Management Group (FIRM) was started by Thorry Gunnersen and other Victorian sawmillers whose resources were threatened by the alpine national park proposal.[1] FIRM publicised how the forest regenerated after cutting and the benefits of the industry to timber towns. It occasionally engaged academics to speak to its case, but without industry-wide financial support it faded away.

In the 1980s the environment movement enlarged its claims to whole categories of forest, such as wilderness and rainforest, and the Australian Conservation Foundation (ACF) advocated as its first forest policy objective in 1983 that 'wood production should be transferred from native forests to ... plantations established outside the current forest estate'.[2] Moreover, the Franklin Dam had shown the powerful influence of public opinion and ability of the Commonwealth to become involved.

The industries needed a more public face and a national approach to defend themselves, but were not well placed because their many organisations were based on States, products or species, while the biggest companies pushed their own interests individually. They did not overcome these difficulties until long after the major environmental organisations were firmly established (Tables 6.2 and 8.1). The pulp and paper companies did not form their national association until 1981, and the main State sawmillers' associations did not embrace other wood industries to become forest products associations until 1983. It was the formation of the Commonwealth's Manufacturing Industries Council in 1986 which threw the industries into working together nationally, and they were eventually persuaded to start national organisations to joust in the national political and public arenas.

The National Association of Forest Industries (NAFI) was at last created in 1987 – twenty-two years after the ACF – to speak for the forest products industries with a single voice for the first time. It was primarily a political lobby group to counter what it saw as 'the major threat ... from the radical environment movement to remove native hardwood forests from production and restrict softwood plantation development'.[3] It also acted as the peak body for several trade and technical matters, and was closely associated with the Forest Industries Campaign Association (FICA) and the Forest Protection Society. FICA was formed under Thorry Gunnersen's leadership slightly before NAFI to conduct a three-year public relations campaign. It spanned States and products, and allowed the ACTU and major industry unions to participate. With a $5.5 million budget for its first two years, it moved the industry's public relations from defensive championing of benefits to attack on the environmental movement.[4] It continued to be funded by the industry after the initial three-year period and in 1993 was absorbed into NAFI.

The Forest Protection Society was formed in Victoria in 1987 as a 'grassroots' organisation to involve people in working communities threatened by the loss of resources for their timber industry. It grew to have a national coordinator, Robyn Loydell, working out of NAFI's office, five State coordinators, and thirty-six groups spread around the timber regions. Their strength waxed and waned with issues and personalities and, much like the environmental groups, they organised local protests and demonstrations, put out newsletters, wrote letters to politicians, and made submissions to inquiries and Environmental Impact Assessments. Various local defence groups formed over particular issues or areas, such as the Bombala Timber Support Group and the World Heritage Opposition Group in Far North Queensland. Pleasingly, the Ladies for Environmental Awareness in Forests (LEAF) emerged from the Timber Workers in Grafton (TWIG)! Many of these groups are part of the Forest Protection Society network. Typically, women take a significant part in their activities, and some of the groups were started by women who realised that:

> by the time the men got home [from logging] they had worked 14–16 hours and they were too tired and worn out to think, ... so the women decided that we'd get organised and we would protect our industry.[5]

At the local government level, thirty-seven Victorian municipalities, led by the Shire of Orbost, formed the Timber Towns Association in 1984 to lobby their State government. The central concern of all these defensive organisations was to stop their local mills being forced to close by the transfer of old-growth forests to conservation reserves.

Mature timber and old-growth forest

The diminishing areas of unlogged oldgrowth forests were both the most important resource of mature hardwood sawlogs and the most ecologically important part of every forest type. The hardwood sawmills were squeezed by the combined effects of reductions in log quotas, due to past overcutting, and mounting environmental claims to the best of their remaining resources. Here, four typical contests are described.

Victorian Alps

One of the earliest contests was over forests in the Victorian Alps. The Victorian National Parks Association (VNPA) had long wanted a large national park in the alpine region, but it was not until 1969 that it felt able to press boldly for a park of 389 000 hectares stretching some 250 kilometres east along the Great Dividing Range from Mt Baw Baw to join Kosciusko National Park at the border.[6] The area contained almost 150 000 hectares of mature alpine ash forest with almost 5 million cubic metres of sawlogs, which supported sawmills in Heyfield, Mansfield and several smaller centres. They had moved to the region after the 1939 fires in the Central Highlands with the hope that they could go back again once the regrowth on the burnt areas – a second forest – was old

Table 8.1 Date of formation of non-government organisations concerned with forests and forest industries

Year	*Base*	*Organisation*
1895	WA	Timber Merchants and Mill Owners Association (lapsed later)
1905	NSW	Associated Country Sawmillers
1913	WA	Sawmillers Association
1928	Tasmania	Tasmanian Timber Organisation
1935	National	Institute of Foresters of Australia
1938	Queensland	Queensland Timber Industry Stabilisation Board
1943	NSW, Vic., Qld, Tas.	Eastern States Timber Industry Stabilisation Conference (ESTIS)
1945	Tasmania	Tasmanian Timber Association (from Tasmanian Timber Organisation)
1947	Victoria	Country Sawmillers Association
1951	WA	Associated Sawmillers and Timber Merchants Association (from Sawmillers Association)
1957	Victoria	Victoria Sawmillers Association (from Country Sawmillers Association)
1959	National	Australian Timber Industry Stabilisation Conference (AusTIS from ESTIS)
1967	Qld	Queensland Timber Board (from Queensland Timber Industry Stabilisation Board)
1969	National	Australian Forest Development Institute (AFDI)
1972	WA	Forest Products Association WA (from Associated Sawmillers and Timber Merchants)
1975	National	Forest Industries Resource Management Group (FIRM)
		Australian Particleboard Manufacturers Association
1978	Tasmania	Tasmanian Logging Association
1981	National	Pulp and Paper Makers Federation of Australia
1983	National	Australian Logging Council
	Tasmania	Forest Industries Association of Tasmania (from Tasmanian Timber Association)
	WA	Forest Industries Federation of WA
	Victoria	East Gippsland Sawmilling Group
1984	Victoria	Timber Towns Association
1986	National	Forest Industries Campaign Association
1987	National	National Association of Forest Industries
1988	National	Forest Protection Society
1989	Victoria	Victorian Association of Forest Industries (from Victorian Sawmillers Association)
1990	NSW	NSW Logging Association
1991	Tasmania	Tasmanian Country Sawmillers Federation

Note: Excludes product development and promotional organisations.

enough to start felling.[7] By the 1970s they had an average of sixteen to seventeen years logging left until they would have felled all the mature, old-growth stands. But the area was also claimed for a large alpine national park, which prompted the Land Conservation Council (LCC) to start studying it. FIRM, the Victorian National Parks Association and many environmentalists commissioned reports and made submissions to the LCC which, in 1978, recommended that 30 per cent of the mature forests should be put in new national parks and 52 per cent should be used for future wood production. Of the remainder, it allowed a small area to be logged once and placed 13 per cent in an 'uncommitted' category for future decision. After more submissions, protests and lobbying, the Victorian government declared the parks as recommended, the mill quotas were reduced, one closed, and the Forests Commission moved another to East Gippsland.

The claims for a great alpine national park had not been fully satisfied. In 1982 a new Labor government with a stronger environmental policy increased the parks by adding the uncommitted land and some other portions. The effects varied across the region. The eight mills in and around Mansfield, for example, agreed to reduce their log allocations in 1979, and then experienced a further cut two years later which left them with only about 30 per cent of their original quota.[8]

East Gippsland

The forests east of the Nowa Nowa and Snowy Rivers had also provided a welcome if remote resource for the sawmilling industry displaced by the 1939 bushfires.[9] By the 1970s there were forty-three sawmills based in sixteen settlements through the region, making the Shire of Orbost more dependent on the timber industry than any other in Victoria. It was also the least modified part of the State as 88 per cent of its area was still public land, mostly forested with some of it still wilderness and with Victoria's largest patch of rainforest at Errinundra. However, logging was pushing further and further into the hills.

After the LCC had studied East Gippsland, four national parks and various small conservation reserves were declared in 1978 and 1979; some areas were left as 'uncommitted' until the next review, and the Forests Commission declared the 1600 hectares of rainforest at Errinundra as a flora reserve. None of these actions removed large quantities of resources from those available to the industry. More serious for the regional economy was the problem that the mills were cutting at a far greater rate than the resource could support and would have to close down in fifteen to twenty years. Only by drastically reducing the log allocations could their life, and hence some level of employment, be maintained. This long-ignored, State-wide problem whose roots lay in devastation by the 1939 fires and the relocation on the industry in the 1950s was included in a public inquiry conducted by Professor Ferguson in 1984–85.[10] After much debate, the government settled on reducing the log allocations from 370 000 to 260 000 cubic metres a year.[11]

At the same time as the mills were trying to cope with these reductions, the East Gippsland Coalition of environmental groups was making claims for 178 000 hectares – of which 175 000 hectares were in forest types with significant timber resources – to be added to the national parks.[12] The new State Labor government ordered the LCC to review the area again, as it had with the Alps, and while it did so placed a moratorium on logging one of the major timber resources with high environmental values, the Roger River Block. The East Gippsland Coalition launched a major campaign to press their case, the National Trust proclaimed its value, and the areas were nominated for listing on the Register of the National Estate. Then, in January 1984, a blockade by the Nomadic Action Group and Forest People near the Errinundra flora reserve turned the coalition's campaign into a confrontation.[13] The government introduced regulations that prohibited obstructing forest operations, and the police arrested twenty-eight protesters.[14] Logging equipment was sabotaged at Errinundra and Bonang, local feelings ran hot, Coalition representatives were jeered from a public meeting in Orbost, and a counter-demonstration by 2500 timber workers from all over the State on the steps of State Parliament called for the moratorium on the Roger River Block to be lifted.[15]

The serious decline in employment drove government agencies and environmental organisations to seek alternatives. The Conservation Council (CCV) published a study which proposed that the reduced quantity of hardwood logs should be cut less wastefully, the timber should be processed much further in the region, logging and forestry should be more intensive, and the national parks should be increased and maintained.[16] Although the possibility of cutting pulpwood was rejected on environmental grounds, the proposed measures were estimated to create about 300 jobs in the forest sector, and an expansion of tourism might add a further 200. If true, it seemed a hopeful prospect.

The LCC then commissioned a rigorous analysis to project the likely economic and employment effects of both its recommendations and the Conservation Council's alternative strategy.[17] In doing this, the analysis recognised that employment could be expected to fall slightly through gradual improvements to productivity. But, most obviously, if the allocations were not reduced, the sawmills would inexorably run out of logs after fifteen or twenty years, and they would provide no employment. If allocations were reduced as the government intended, employment would fall by two-thirds (Figure 8.1). If in addition more parks were declared as the LCC recommended, then employment would fall even further (Figure 8.2, Cases A and B). But irrespective of the parks, the most significant option was to introduce an integrated regime to cut both sawlogs and pulpwood. This could work over previously logged areas to produce 585 000 cubic metres of pulpwood a year (Case D), or be restricted to new areas and produce only 125 000 cubic metres a year (Case C). The environmentalists' proposals for adding value to the timber by further processing were estimated to add about fifty jobs, but the expansion of tourism was found likely to be little affected by the forest decisions. From this analysis, it was apparent that the most hopeful prospect for employment was to cut pulpwood for export as woodchips or to supply a new pulp mill. The environmental movement strongly opposed both these options.

As the LCC and Ferguson inquiries dragged on, the prospects for the mills, workers, towns and settlements became gloomier and gloomier. To gain media and political attention, the loggers and mill workers took the drastic action of blockading the Princes Highway near Orbost with sixty log trucks.[18] It was to little avail. The LCC recommended that forests containing one-third of the remaining sawlogs should be placed in further national parks (Table 8.2). This, and revisions to the calculations of how much could be cut, left the sawmilling industry with only about 100 000 cubic metres a year, an overall reduction of two-thirds from the original quotas. Various possibilities of increase were explored, but the chance of obtaining sawlogs from some of the uneconomic areas by also cutting some pulpwood for woodchip exports was precluded by a government moratorium. Another possibility was to build a pulp mill, but this too came to nought, as discussed later.

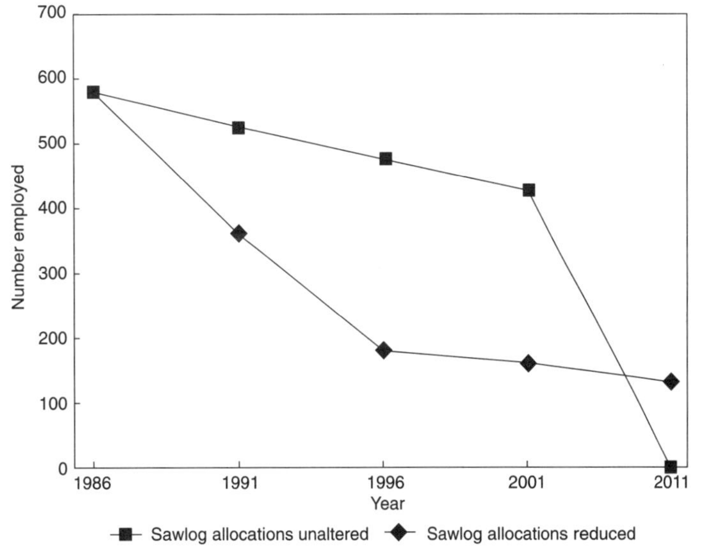

Figure 8.1 East Gippsland timber employment projections – sawlog allocation options
Source: Stanley and King 1986: 102.

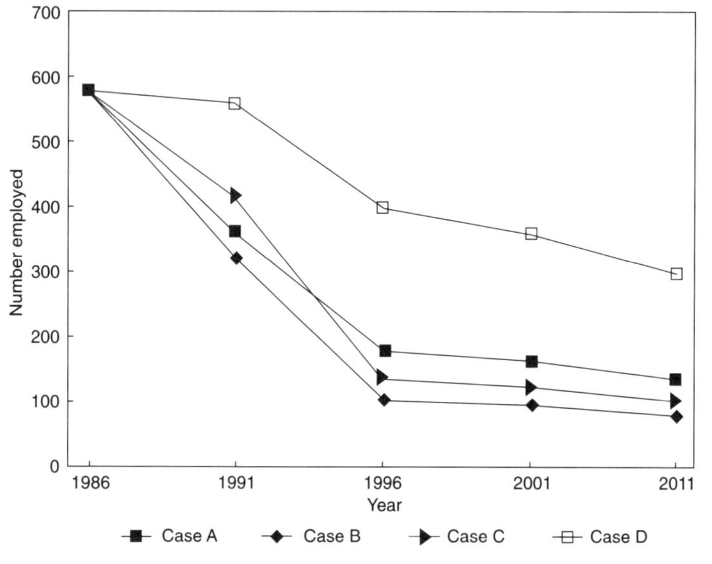

Figure 8.2 East Gippsland timber employment projections – parks and pulpwood options

Case	*Parks*	*Pulpwood*
A	No more	None
B	More	None
C	"	High
D	"	Low

Source: Stanley and King 1986: 102.

Table 8.2 Sawlog resources in East Gippsland, 1985

		Volume of sawlogs (cubic metres)
Estimated sawlog timber resources:		
on all public land in East Gippsland		12 368 000
Less:		
in areas uneconomic to harvest and/or regenerate	5 069 000	
in conservation reserves created prior to the 1980s	763 000	5 832 000
Available for harvesting prior to 1986		6 536 000
Less resources in proposed new parks:		
Roger River–Bowen Range	1 270 000	
Errinundra	754 000	
Coopracambra	176 000	2 200 000
Available for harvesting after new parks		4 336 000

Source: LCC (1986).

Note: Excluding special uncommitted area subject to review.

The environmentalists were still not satisfied. The LCC had recommended only 109 000 hectares of the 175 000 they sought and not quite all of the area on the Heritage Commission's interim list; some old-growth areas were still to be logged. Under pressure from the Commonwealth as well, the Victorian government agreed to a temporary moratorium in 1990 while the effect on resources was investigated yet again. A further 6500 hectares was put into national parks, logging was eventually resumed in much of the remainder, detailed management plans were commenced, and the government cooperated with the Heritage Commission in a joint study. However, environmental objections and protests continued as the East Gippsland Coalition was replaced by the more radical East Gippsland Forest Alliance, which, with the Friends of the Earth, ran further blockades and demonstrations in 1993 and 1994.[19] The atmosphere of uncertainty continued to hang over the industry and the livelihoods of the people working in it.

Ash regrowth forests of the Central Highlands

The future of the Alpine and East Gippsland mills was always thought to lie in moving back to the second forest growing in the Central Highlands after the 1939 fires, but as time passed it appeared less certain. The main species – mountain and alpine ash – need 80 to 120 years to grow the prime seasoning grade timber for which they are prized most; but the mills started to return in 1980s and 1990s as the East Gippsland resource declined. Although there were some trees of 60 to 70 years from earlier regrowth, the main resource from the 1939 fires was only 40 to 50 years old, and it proved more difficult to saw and the timber was of poorer quality than expected. In 1983 another major fire burnt 45 000 hectares in some of the most productive of the mountain ash forests.[20] Although as much timber as possible was salvaged, the loss put even more pressure on the remaining resource. Moreover, the plan to cut through the forests as the trees came of commercial age would not enable the second forest ever to grow really old trees – of 200 to 300 years – which produce the best habitat for creatures such as Leadbeater's possum; the felling plans and patterning of the forests would have to be altered to accommodate them.[21] On top of this, any rapid felling of the second forest would diminish the amount of water which could be collected from the catchments and supplied to an increasingly thirsty Melbourne.[22] All told, the original intention would have to be considerably modified; the regrowth forests had acquired new values in their potential to become old.

Far north coast of New South Wales

The 1982 decision to preserve the rainforests in northern New South Wales transferred 104 000 hectares from the Forestry Commission to the National Parks and Wildlife Service, thus effectively removing 435 000 cubic metres of logs, mostly hardwood, from the total resources available in the region.[23] It came on top of reductions of mill quotas by 38 per cent between 1980 and 1990 which forced three mills, including a ply mill, to close.[24] The Forestry Commission managed to buy a eucalypt plantation from Amcor and nearly 12 000 hectares of forests from a sawmilling company, but these new resources did not fully compensate for the effects of the rainforest decision.

The environmental claims moved on from the rainforests to the surrounding hardwood forests. These too were biologically profuse, a 'veritable forest-dependent zoo, probably unparalleled in south-eastern Australia'.[25] They were also resources of high-quality eucalypt, brush box and tallowwood sawlogs, some areas of which in rugged country were still being opened up for logging. The resulting contests came to a highly publicised climax over an area of 560 hectares in three compartments of the Chaelundi State Forest.[26] They were adjacent to the Guy Fawkes River National Park and the area had been nominated, but not then assessed, as being suitable for reservation as a wilderness.

Several hundred timber workers and their families from northern New South Wales demonstrated opposite the Glen Innes town hall on 6 May 1981 while the Cabinet considered whether the Washpool forest should be a national park.
(NSW Forest Products Association, *Daily Examiner*, Grafton)

Two thousand timber workers and their families from all over NSW protested outside State Parliament on 10 March 1992 while an emergency session debated whether to allow logging in fifteen areas while EISs were prepared. (Anthony Farr)

It contained one of the last undisturbed stands of tall old-growth rich in tallowwood.

When the Forestry Commission started to drive a new road into the area in 1990, the North-East Forest Alliance (NEFA) organised a blockade and one of its members, John Corkill, obtained an injunction from the Land and Environment Court which prevented further work until an EIS was prepared. A few months later the commission issued it, then approved it in 1991. A second blockade, during which more than 200 people were arrested, managed to stop logging for four months. It had hardly started again when John Corkill obtained a judgement, upheld on appeal, that recognised the obvious dependence of creatures on their habitat for food and shelter.[27] Although the Forestry Commission had prescribed some measures to ameliorate the impacts of logging on wildlife, it had not sought permits from the State's National Parks and Wildlife service to 'take, kill, disturb or injure' any animals. The court found that roading and logging would disturb or injure twenty-three species, some of which were endangered. Logging could not proceed. After this experience, the State introduced new legislation which required every coupe to be checked for endangered species. If any were present, permits had to be obtained from the Parks and Wildlife Service before any logging could start.

Second-order contests

Major confrontations, such as those at Chaelundi, or Terania Creek a decade earlier, were highly publicised in State and national media. But there were many lesser, 'second-order' contests and local conflicts throughout Australia which were reported only in the local newspapers.

Table 8.3 Classification of reported events relating to forests in northern New South Wales, 1983–94

Event	*Number*	*Proportion (%)*
Pro-logging		
Objection, complaint, etc.	54	14.6
Meeting	26	7.0
Court/legal action	18	4.9
Government decision	11	3.0
Uniting strategy, future plans	11	3.0
Reports, publicity	11	3.0
Demonstration	9	2.4
Industry cost cutting	3	0.8
Threats	1	0.3
Anti-logging		
Objection, complaint, etc.	52	14.1
Court, legal action	40	10.8
Blockade, demonstration	36	9.8
Meetings	10	2.8
Arson, injury, threats	8	2.2
Submissions	7	1.9
Government decision	6	1.6
Damage to roads	6	1.6
Uniting strategy, future plans	3	0.8
Publicity	3	0.8
Pro-pulp mill		
Objection	1	0.3
Other	3	0.8
Anti-pulp mill		
Objection	16	4.3
Meeting	4	1.1
Other	5	1.4
Other	25	6.8
Total	369	100.0

To canvass them in every State would be wearisome and beyond the scope of this book. Nevertheless, their variety and recurrence affected people's lives and livelihoods. The situation in northern New South Wales can be glimpsed from a classification of reports in fifty local newspapers in the region between 1983 and 1994 (Table 8.3).[28] Distinct 'events' were classified according to whether the 'pro-logging' side was promoting them, the 'anti-logging' side, or an 'anti-pulp mill' side which was not necessarily anti-logging.

To the sawmillers and loggers, there seemed to be an endless succession of claims, blockades, EIAs, court actions and bureaucratic fights raging up and down the coastal forests, which diminished the resources available and caused them endless delays and disruptions. They reacted in several ways. Politically, the NSW Forest Products Association urged their case to government and flooded the press with statements. Locally, they formed Forest Protection Society groups and organised their own demonstrations in local towns. But many were so disheartened or defeated by the conflicts that they simply left the business.

'Old-growth' as an icon

Such contests turned 'old-growth' into an icon to be claimed. Like 'wilderness' and 'rainforest' before it, the contestants disputed its definition once the claim took hold. All included ecologically mature forest with negligible disturbance, but some would accept younger ages and others more disturbance.[29] But how much was ecologically mature, how much undisturbed? By the 1990s it was estimated that overall about one-fifth of Australia's state forests and two-thirds of its conservation reserves had never been logged.[30] However, in some of the most productive types of forest the proportions were far less and the unlogged area was expected to decrease by between 12 and 30 per cent if logging trends continued. Clearly these were going to be subjects for continuing the major contests unless policies or attitudes changed.

Social and regional impacts

The industrialists' greatest political weapon was the threat of unemployment. Conveniently disregarding the unemployment they created through structural change and mechanisation, they wielded it with a will against the environmentalists. For example, FICA applied very 'creative algebra' to forecast that the ACF's policies would result in the loss of up to 86 700 jobs; but the sober RAC calculated 16 400.[31]

FICA advertised nation-wide that 'Australia's managed forests are the lifeblood of [85 listed] towns ... all in danger.'[32] But the forest industries are only a small component in most towns, and ten of those that FICA listed relied on plantations anyway. While the industry associations exaggerated and scared, some environmental organisations proposed alternatives, and the Commonwealth offered compensation.

Alternative employment strategies

The environmentalists offered ideas. Their proposals were made during the 1980s when a 'bottom-up' approach to combat unemployment by encouraging the formation of small, local enterprises was widespread.[33] However, these proposals were designed to offset the loss of jobs if more conservation reserves were created rather than to counter existing unemployment. They made the environmental claims more socially and politically acceptable, and to their authors were a responsible way of advancing them.

Tasmania

The first major strategy was prepared for the Tasmanian Conservation Trust to provide an alternative to exporting woodchips from Tasmania after 1988.[34] It aimed to maintain employment without draining public finances, while also increasing the reserves and protecting the environment. It put up suggestions for more efficient use and further processing of the hardwood logs sawn in Tasmania, raising design skills in the furniture industry, and using waste wood for industrial fuel. To continue wood production from the smaller area, it proposed that the regrowth stands should be thinned and plantations should grow pulpwood. It also proposed to change the integrated concession regime to a quota system, increase the use of small wood, and ensure that more of the logs being cut in the Australian Newsprint Mills (ANM) concession were diverted into sawmilling or veneer production. Overall, it hoped that an 'industry plan' could be negotiated, as had been done for the steel industry, which would maintain employment from fewer resources.

The woodchip companies and Forestry Commission retorted that many of the suggestions had been commenced already. They rejected the reservation of any additional areas of production forest, and they argued that any benefits flowing from the suggestions should be additional to the existing industry, not in place of it.[35] Woodchipping continued, but most of the other suggestions were eventually incorporated in a Forestry and Forest Industry Strategy. The Tasmanian initiative in devising alternative paths for regional development was followed in Victoria, as mentioned earlier, and in south-east New South Wales. Their central thrust was to add value to whatever timber was cut by processing it further, especially in small-scale, local industries.

Adding value

Victoria followed up the Ferguson inquiry by preparing a detailed strategy for the timber industry, key features of which were a comprehensive planning system and a Value Adding Utilisation Scheme (VAUS).[36] This altered the way in which logs were allocated to sawmills. Instead of mixed lots of logs being sent to each mill from its specific area, the new arrangement sorted the logs in the forest so that the highest-quality logs were sent only to mills equipped to produce fine timber from them, and the lower qualities of logs were sent to mills producing lower grades of products, such as green scantling, fence palings or pallets. In spite of difficulties, the scheme facilitated investment and the development of some new products. For example, it enabled logs from the mixed forests in the goldfields area north of Melbourne to be sorted by species so that the Black Forest Sawmills could obtain all the messmate logs. CSIRO was then able to develop kiln-drying schedules specifically for that species so that the mill could process and market messmate as an identified, premium product.

The Commonwealth referred the potential for adding further value to Australia's forest products to its Industry Commission. Not surprisingly, this hard-line body of economic rationalists issued a report recommending the removal of all brakes on market forces. It specifically recommended that 'Governments cease to use log royalties, log allocation procedures or any other instrument to direct output of the forest product industries towards the production of higher value added products.'[37] An internationally competitive market would do it better, or so it believed.

Compensation

When the Commonwealth had won the various constitutional contests with the States and could force its will, it was left with the political problems of unemployment and re-establishing relationships with the States who still had to manage the land. Its solution was to offer *ex gratia* compensation packages to the States affected (Table 8.4).

North Queensland and Fraser Island

The declaration of the North Queensland rainforests as World Heritage in 1988 affected the livelihoods of timber workers, particularly in Ravenshoe, Millaa Millaa and Yungaburra. It was thought that quota

Table 8.4 Examples of Commonwealth compensation payments for natural resource decisions

State	*Contest*	*Amount ($ million)*	*Payee*
Qld	Fraser Island sand mining	10.9	State, companies
Tas.	Franklin Dam	800	State, companies, individuals
Qld	Rainforests	75	Companies, individuals
Tas.	Lemonthyme and Southern Forests	50	State
Vic.	East Gippsland National Estate forests	10	State
Qld	Fraser Island forests	16.5	State, individuals

Source: RAC (1991), vol. 1, p. 567.

reductions and mill amalgamations would reduce employment by 10 per cent anyway, and that the effect of the World Heritage listing would be to reduce employment from the remaining 800 to 200, clearly a drastic effect on these small towns.[38] The Commonwealth developed a Structural Adjustment Package which distributed $75 million to the sawmillers, loggers and workers, and to alternative employment projects, mostly in the public sector. Unemployment occurred much as expected, but the compensation did not fully meet the needs. A dislocation allowance of $2500 to timber workers who had lost their jobs helped only in the short-term; an early retirement scheme for those between 55 and 60 provided much less than they could have earnt otherwise; and the public works projects had finite lives, after which little other work was available in the region. When Fraser Island was listed as World Heritage in 1992, the Commonwealth provided $16.5 million to which the State government added $21.2 million for regional development.[39] Part of this was used to provide cash compensation of between $5000 and $40 000 for each of the timber workers who lost a job.

Other States

The divided report of the Helsham inquiry into Tasmania's Lemonthyme and Southern Forests left the decision over World Heritage nomination to the Commonwealth. When it finally decided in favour of nominating the larger area, it provided Tasmania with $50 million to be used largely for creating alternative resources by managing other forests more intensively and by establishing plantations. In East Gippsland, the Commonwealth provided $10 million to the Victorian government in compensation for its agreement not to log some of the National Estate areas outside the extensive park system that had been created.

These various schemes reacted to, and at best ameliorated, the social impacts of the environmental claims. But industry did not remain simply reactive; it too made claims and took the initiative.

Capital's initiative

To counter the general malaise of Australia's manufacturing, the Commonwealth set up a 'tripartite' Manufacturing Industries Council of government, industrialists and unions in 1984 to see what could be done. One of its divisions, the Forestry and Forest Products Industries Council (FAFPIC), was made up of Commonwealth officials, the heads of the State forest services, leaders from each of the major forest industries, and union officials. It was a powerful assembly which, in spite of its disparate interests, was united in its opposition to environmentalism and its belief in expansion.

It took stock of the industries' position, concentrating on the uncertainty of future markets and 'the need to prevent unreasonable losses of the forest estate on the basis of protection of non wood values'.[40] It noted a 'lack of recognition' of the importance of the industry. It thought that prospects for woodchip exports were bright, 'given a strong, positive attitude by Australian governments ... as licences are renewed', and it raised the possibility of building a world-scale pulp mill when the market recovered. It commissioned a study of international economic competitiveness, which found that the existing sawmills were too small and costly, but that new sawmills should be able to compete with imports.[41] For the pulp and paper industry, it found that domestic production could replace imports if big enough mills could be built. For the export market, producing pulp was thought to be only a first step towards integrated pulp and paper mills which would be need to be large to be competitive – consuming over 1 million cubic metres of wood a year. With this in hand, FAFPIC prepared a 'strategic plan' for high-value wood production and export based on more intense forest management, long-term resource rights for companies in the public forests, cheaper wood, and an end to transferring wood resources to conservation reserves.[42]

FAFPIC's 'Growth Plan'

Although the study of competitiveness was guarded in its assessments, FAFPIC was not. Its manifesto for the period to 2030, the *Forest Industries Growth Plan*, proffered $11 billion of investment to build three new hardwood pulp mills, primarily for export, six softwood pulp and paper mills, and a string of sawmills and panel mills – all of which were to be of an internationally competitive 'world-scale'.[43] This, it asserted, would reverse Australia's import bill for forest products and create 125 000 new jobs. It would need all the resources from the native forests and another 600 000 hectares of plantations – mostly pine – to be established at a cost of $2.1 billion.

ACF's conservation strategy

The ACF took a different view in its own manifesto, *The wood and the trees.*[44] It proffered a 'sustainable and sympathetic relationship between the human economy and the natural environment' for the forest sector by establishing eucalypt plantations which could replace the hardwood resources of the native forests over 15–30 years. The native forests should be in national parks, except for a few forests managed on cutting cycles 'of the order of several centuries' to provide small volumes of high-quality logs for making furniture. It called not for exporting woodchips or pulp, but for domestic manufacturing and more recycling of waste paper. The reduced demands for wood could be met by continuing the existing rate of planting pines and increasing the rate of planting eucalypts to add 300 000 hectares of eucalypt plantations to the total estate. Comparatively, it was a modest program which initially seemed to offer the entrancing prospect of ending the environmental contests. But FAFPIC quickly claimed that any hardwood plantations should supplement, rather than replace, the resources of the native forests.[45]

Expanding the plantations

To their different ends, both manifestos called for more plantations, but was there enough land? would eucalypts grow well? and who would pay?[46] The States' forest services were embroiled in a wave of corporatisation and privatisation, so that public funding was not likely to be the source for major expansion; it would have to be left to private investment decisions.

Large industrial companies had already developed almost one-third of Australia's plantations. Their 260 000 hectares of pine plantations produced both pulpwood and sawlogs, but their 50 000 hectares of eucalypt plantations were intended only for pulpwood as it seemed that the longer rotations required for high-quality eucalypt sawlogs were not economically attractive.[47] Individual farmers and investors had established only about 6 per cent of Australia's plantations, aided by various extension and subsidy schemes. They experienced marketing, taxation, price and other difficulties, but the expansion of this resource was attractive for several reasons.[48] First, it would avoid the community objections to state purchase of agricultural land for its plantations; in Victoria objectors had even got to the point of pulling pine seedlings out of the ground and blockading roads.[49] Second, it offered an opportunity to diversify farm production and incomes; and third, it could be important in catchment management and countering land degradation. These last were important in a National Afforestation Program which funded the establishment of 6000 hectares of hardwood plantations between 1987 and 1992 and supported research into their growth.

The Commonwealth, the States and FAFPIC all hunted for information, and a National Plantations Advisory Committee was set up in 1990 to see what could be done through private investment 'where commercially viable'.[50] It commissioned studies, found that there was about 500 000 hectares of suitable land, and recommended that the Commonwealth change the tax arrangements to allow for the long period of plantation investments. Legal changes were needed, too, for joint venture schemes. Beyond that, it thought that better information and regional consultative committees would make the market system work better for small growers. Detailed economic studies followed, confirming that plantations were likely to be an internationally competitive means of providing softwood to sawmills and particle-board factories, and hardwood to pulp and paper mills.[51]

Even the best reports could not plant trees. That needed capital, and in Western Australia it came from Japan and Korea to plant eucalypts on farmland. Western Australia had long been planting pines and in the 1980s planted blue gum to 'rehabilitate' bauxite mine sites. They grew so well that they offered a remedy for the State's great problem of increasing salinity in the water table. Trees had to be put back on farmland that should never have been cleared, and the Department of Conservation and Land Management (CALM) sought commercially viable ways for hard-pressed land-owners to do so. Farmers needed income while the tree crops were growing, and often some work too; the forest service needed land without having to buy large tracts; and Japanese companies needed long-term supplies of high-quality pulpwood. The solution was for CALM to negotiate agreements with both foreign companies and Australian farmers. The first agreement in 1993 was made with the Japanese Oji Paper Company and a sogo shoshu, Itochu Corporation, who expected to invest $60 million over ten years. The agreement provided for CALM to locate suitable land, arrange sharefarming agreements with the land-owners, establish the plantations and then manage them on behalf of the investors. It envisaged planting 20 000 hectares of blue gums in lots of 20 to 100 hectares on different properties around the port of Albany. The trees are to be harvested as pulpwood at ten years of age and then allowed to coppice to produce a second crop ten years later, all of which would be exported to Japan as woodchips.[52] A second, similar agreement was made with the Korean Hansol Forest Products Company to establish 15 000 hectares of plantations over ten years near the port of Bunbury.

Expanding the industry

Sawmills, panel factories, tissue mills and a new pulp mill at Maryvale which drew on the fattening pine resources for domestic markets were built or enlarged during the 1980s with little or no public controversy. But proposals to build two large mills which would draw on the hardwood forests to produce pulp for export, and some replacement of imports – the leading feature of FAFPIC's manifesto – were fiercely opposed, and for the time being defeated. One was at Wesley Vale in Tasmania, the other in East Gippsland.

Wesley Vale

As mentioned in Chapter 4, North Broken Hill (NBH) made the first proposal to build a bleached kraft export pulp mill in 1988. APPM, by then its subsidiary, had talked about it for years and had gained control of three-quarters of Tasmania's pulpwood resources. When the Japanese pulp market seemed to be improving and the Australian dollar was depreciating, a new mill at last seemed possible. NBH proposed to build it adjacent to its existing paper mill at Wesley Vale in northern Tasmania.[53] The mill would make 440 000 tonnes a year of pulp, mainly for export, from 1.7 million tonnes a year of hardwood being exported as woodchips or used at the company's old Burnie pulp mill. This would leave a significant quantity which, the company thought, could still be exported as woodchips. The proposed investment of, it was said, $1 billion delighted the Tasmanian Liberal government; upgrading woodchip exports to a higher-value product seemed just what everybody wanted; no more wood was going to be cut, and there would be more jobs.

It was a surprising proposal, because Japanese companies had already built pulp mills to process the Tasmanian woodchips in Japan. In 1987 it was reported that a group of them had committed $100 million to the $1 billion project, but it is unclear how firm this was. In

Table 8.5 Chronology of Wesley Vale pulp mill contest

Year	*Event*
1961	Wesley Vale concession granted to APPM.
1970	APPM builds paper mill but defers pulp mill.
1972	APPM exports woodchips from Wesley Vale concession.
1979	APPM buys TPFH woodchip mill.
1983	APPM taken over by NBH.
1987	NBH decides to develop pulp mill at Wesley Vale and allocates $2 million for project planning.
	Japanese investment reported.
1988	CROPS group formed and opposes siting and pollution.
	NBH releases Environmental Review statement.
	NBH merges with Peko-Wallsend which increases its financial strength.
	Project to be joint venture with Noranda.
	NBH threatens Tasmanian Government that project will be stopped unless it approves the Wesley Vale site.
	Tasmanian Cabinet approves site.
	Dispute over royalty rates. NBH again threatens to withdraw and Government reduces rates slightly.
	Commonwealth announces incentives for the mill.
	Fishing Industry Council opposes mill.
	NBH and Noranda announce they will proceed and request fast tracking of approvals.
	EIS issued and queried by Tasmanian Minister for Environment.
	Northern Pulp Mill Agreement Act 1988 (Tas.) passed.
	Addendum to EIS issued.
	Environmental guidelines prepared.
1989	NBH and Government negotiate over environmental guidelines which are then passed by Tasmanian Parliament.
	NBH threatens to end agreement unless environmental guidelines 'reconsidered' and approved. Guidelines changed.
	Commonwealth Government calls on CSIRO for scientific report.
	Noranda threatens to pull out unless State and Commonwealth approvals granted quickly.
	CSIRO study reports 'inadequacies' of 'environmental justifications' in EIS.
	Federal Cabinet decides that new environmental guidelines will be prepared.
	Noranda withdraws and project lapses.
	Tasmanian Liberal party loses Government and five independent Green candidates elected.
1993	NBH sells its pulp and paper mills to Amcor.

Source: Milne (1990) and other sources.

any event, the venture took a new structure when a Canadian company, Noranda Forest Inc., took a half-share in the project, with arrangements for the Japanese market to take 'a large proportion' of the pulp. Noranda, with several mills across Canada, was four times the size of NBH in the pulp and paper industry. It said it was restricted for expansion in Canada and that it could use some of the Tasmanian pulp in its paper mills in British Columbia.[54]

Complicating the Tasmanian proposal was a growing realisation that chlorine bleaching in kraft pulp mills produced some organochlorine compounds – dioxins and furans – which even in minute quantities were very harmful to human health. It was an international problem. In Scandinavia some mills had moved to oxygen delignification to reduce the problem, while emissions from some North American mills became a public issue there at the very time the Tasmanian mill was being proposed. Industry associations and government regulators surveyed dioxin emissions, their presence in 'feminine personal care' products, their accumulation in lakes and rivers, and how they passed through the food chain to contaminate birds, fish and shellfish near pulp mills. The results were not definitive, but in 1989 the Canadian government, like those of other countries, ordered the pulp mills, including Noranda's, to plan how and when they would reduce the emissions of toxins.

Objections to the Wesley Vale mill were first raised by local residents and farmers who did not want such a large industry to disturb their peaceful agricultural area. It was, they felt, an inappropriate land use. They formed the Concerned Residents Opposing Pulp mill Siting (CROPS) group, led by a local school teacher, Christine Milne, to voice their concerns.[55] Joined later by fishermen, they were a new source of opposition from a constituency which had traditionally supported the Liberals. NBH issued an *Environmental Review* which failed to allay their fears. Meanwhile, as the possible environmental impacts of the project and the North American furore over dioxins became more fully realised, the base of opposition widened. It was no longer a question of where to put the mill, but whether it should go ahead at all.

APPM was used to having its own way with the State government which, up for re-election and desperate for development, did all it could to fast-track the project through the approval process. The companies repeatedly threatened to withdraw if their requirements were not met, and smartly too. They feared that their market might be caught by new Brazilian mills, and Noranda was vulnerable if higher environmental standards were imposed in Tasmania than were being developed in Canada. CROPS certainly tried to embarrass Noranda with a letter-writing campaign to the major newspapers in Toronto, while Greenpeace in Canada campaigned against the pollution standards in Noranda's Canadian mills. Although the Wesley Vale mill was designed with sophisticated emission controls, there was considerable uncertainty about the level of organochlorines which would be emitted, how these would be diffused and what the effects, particularly on marine life, would be. Then followed a farcical period during the summer of 1988–89 when the Tasmanian government, bullied by the companies and harried in the media, vacillated and finally weakened its environmental guidelines. Although the Premier

Robin Gray, Premier of Tasmania, assures Graham Richardson, Federal Minister for the Environment, of the safety of the proposed Wesley Vale pulp mill.
(Geoff Pryor, *Canberra Times*, 24 February 1989)

acknowledged that dioxins would be discharged into Bass Strait, he said that when the United States set a standard, Tasmania would follow it, thus neatly removing the vulnerability of Noranda's mills to higher standards.[56]

However, the Commonwealth Labor government had to approve the project too through its Foreign Investment Review Board. It wanted any development which would reduce unemployment and the adverse balance of payments, and it was reluctant to cross the Tasmanians, having lost seats there after the Franklin dispute. At first it offered a package of financial incentives, including accelerated depreciation, to encourage the development.[57] But when thousands of protesters rallied in Hobart, and the Tasmanian government watered down the environmental guidelines, it could ignore the environmental problems no more. Amidst the claims and counter claims clarity was lacking, so it called on the CSIRO to report on the level of dioxins to be discharged into Bass Strait and the general question of organochlorine pollution.

When the report arrived, it became clear that the mill was proposing to use the most modern commercially proven technology for pulping and bleaching, but that Bass Strait probably had limited ability to disperse the effluent to which the marine organisms were likely to be susceptible.[58] The lack of survey information to form a baseline was crucial, because the companies could only be held responsible under the Tasmanian agreement if pollution could be clearly proved. The federal Cabinet was keen to get the project but insisted that the baseline would have to be established by survey, and that superior standards to be laid down and monitored by the Commonwealth would have to be met.[59] The companies stood to lose the depreciation allowance by delay, and possibly some deposits on machinery orders if finally the mill proved impossible. They did not accept the Commonwealth's ruling that they would have to meet standards which had not yet been set. Noranda withdrew and the proposal lapsed. The State government fell at the election, and Christine Milne, with four other Green Independents, was elected.

East Gippsland

Although some of the forests in East Gippsland were within the economic radius of the woodchip mill at Eden, Victoria husbanded its resources, hoping for development within its borders. It lacked the fine port that Eden had, but the Forests Commission tried to see how else its wood could be exported. By 1975 the ACF and the CCV were opposing woodchipping strongly and, with the spectacle of the early operations at Eden nearby, persuaded Hamer, the Liberal Premier, to halt some trial operations until the LCC had studied the area.[60] When this was done, the Forests Commission tried again in 1982, calling tenders for 500 000 tonnes a year and issuing an EIS, but the new Labor government declared a moratorium before anything else was done. Its Ferguson inquiry (1984–85) and Timber Industry Strategy (1986) both thought that pulpwood could come from areas being cut for sawlogs without harm. The pulpwood – euphemistically renamed 'residual roundwood' – was included as part of the VAUS scheme, and the strategy made it very clear that 'in Victorian forests there will be no pulpwood-only operations'.[61] But opposition to woodchipping was so fierce that even a trial of the VAUS system which involved sending 100 000 tonnes a year for three years – not a big volume for the region – to Eden was halted until an Environmental Effects Statement had been prepared.[62] Then it had to be assessed by an independent committee of experts, which recommended an extension of the period and a careful examination of past cutting on other areas.[63]

Table 8.6 Chronology of East Gippsland pulp mill contest

Year	*Event*
1970	Victoria not a party to Eden (NSW) woodchip export project.
1975	Forests Commission commences environmental study into feasibility of pulpwood industry in East Gippsland. Licence for 120-hectare trial area granted, and then revoked until LCC study completed.
1976	Forests Commission advocates woodchip export schemes for region.
1977	Native Forest Action Council survey reports that less than one-third of Melburnians know of plans for woodchipping in region, and of these most disapprove.
1980	Forests Commission preparing EIS on woodchip industry for region.
1981	Tenders called for 500 000-tonne a year woodchip contract. 100 000-tonne trial approved. EIS issued. Labor Party pledges to stop further woodchipping in hardwood forests.
1982	Labor wins government and declares moratorium on woodchip licences.
1984	Ferguson Inquiry. East Gippsland Coalition formed.
1985	Timber Industry Strategy being negotiated.
1986	Timber Industry Strategy issued allowing 300 000 tonnes of pulpwood a year to be cut for five years, subject to an EIS.
	Some East Gippsland forests put on register on National Estate and Government agrees not to harvest them that year.
	LCC commissions study of economic and employment implications of its recommendations.
1988	100 000 tonnes of pulpwood cut for woodchip exports. Government calls for expressions of interest in feasibility studies for a pulp and paper mill in East Gippsland. APM and NBH are chosen to compete in next stage.
1989	(NBH–Noranda Wesley Vale proposal rejected by Federal Cabinet. Major confrontation over logging in National Estate forests in adjacent Eden region.)
	Government postpones pulp mill proposal indefinitely.
1990	Government decides to add 6500 hectares of National Estate forests to national parks but to allow logging in 74 000 hectares.
1991	Pulpwood trial to go ahead.

Source: Mortlock and Dargavel (1989) and other sources.

In 1988, the government called internationally for expressions of interest from companies prepared to study the feasibility of building a pulp and paper mill. From five respondents, it chose Amcor and NBH to prepare 'competitive pre-feasibility studies' by 30 June 1989, from which it would choose one to proceed to a full study. A complicated approval program was specified, which required the results of the VAUS trial to be evaluated while the mill designs and EIS were being prepared. A maze of government committees sat over this: at the top was a sub-committee of the Economics Committee of Cabinet, then an Inter-departmental Management Coordination Group, then an East Gippsland Paper Products Development Study Project Team which had to work closely with an Inter-departmental Reference Group.[64] The Project Team structure worked well, providing a line of contact for the companies which could overcome inter-departmental differences.

Meanwhile, the Victorian government was being pressed to stop logging in those National Estate areas still outside national parks, and just over the border, a major confrontation over logging in National Estate forests at Tantawanglo and Coolungubra was in full swing. Only two days before the deadline for the competitive studies to be completed, the government indicated that in view of the conflicts with the Commonwealth over the National Estate, it had decided to postpone the whole process. It had not been resumed by 1995.

Approving new mills

The Wesley Vale fiasco showed that better EIS and approval processes were badly needed if new pulp mills were ever to be built.[65] The Commonwealth sent a review team around the world to examine the best technology and collect the latest information on the toxicological effects of mill effluent on marine life. The bleached kraft process was confirmed as the only viable option for Australia's mature eucalypts; if the 'best technology economically achievable' was used, it was not expected to cause significant ecological impacts in the short term, although long-term chronic impacts were always possible. The Commonwealth then issued its guidelines, which set standards for emissions, and stringent criteria for baseline studies and monitoring. It also funded research into improving the technology and understanding the environmental impacts. The guidelines were made subject to review as this research and pulping technology advanced.

Pollution was, of course, a world-wide problem for the pulp industry. As mentioned, several Scandinavian mills changed to oxygen bleaching processes which were totally free of chlorine (TCF) and hence the toxic organochlorine compounds. Most Canadian and United States mills changed to using chlorine dioxide and were thus free of elemental chlorine (ECF). This was expected to reduce the emission of dioxins to undetectable levels.[66] By 1994, about 10 per cent of the world's pulp was produced by the TCF processes and about one-half by the ECF processes. In Australia, a TCF oxygen-based bleaching process was selected in 1993 for an expansion of Amcor's tissue mill in South Australia (half owned with Kimberley–Clark) in 1993.[67]

The first export pulp mill proposal to comply with the Commonwealth's guidelines is likely to be built in Western Australia. In 1992, the Western Australian government called for expressions of interest in ways in which the woodchip export industry could be upgraded. Oceanographic studies were commenced, National Estate values were studied jointly with the Heritage Commission, and Bunnings Ltd, which already owned the woodchip mill, was selected to undertake a feasibility study.[68] It proposed a small chemi-mechanical mill as a first stage to be followed, if successful, by a full kraft mill. The earlier proposals to build mills in northern New South Wales, northern Tasmania (Wesley Vale) and East Gippsland might also be reactivated, but it seems that they will need to be supported by plantations as well as the native forests. Bunnings' proposal, for example, included expanding the area of its blue gum plantations to more than 30 000 hectares.

Continuing expansion

From the industrialists' perspective, the environmental contests over the hardwood forests appeared disastrous. Their developments were no longer regarded as unambiguously worthy by the public mind, nor perhaps by institutions which might finance them. Their resource regimes lost the expectation – entrenched longer in administrative practice than in property rights – of being able to log virtually every tree. National parks were no longer created mostly in economically residual lands,[69] creeks and steep slopes could not be logged without regard for erosion, nor the havens for birds and beasts destroyed. Nor were the States the sole arbiters of forest use and management on their behalf: the Commonwealth also took a role. Would the environmentalists hound industry from the native forests? Would impossible standards be set? Would proposals stagnate while the States and Commonwealth squabbled? These new uncertainties were felt – and decried – as barriers to investment. Somehow, governments had to restore investment confidence, while answering the environmental claims and dealing with the social and regional impacts of restructuring. The following chapter describes the political dimensions of their attempts to do so.

CHAPTER NINE

CONTESTING POLITICS AND THE STATE

The claims on Australia's forests were contested in the arenas of politics and the state, changing both the forests and the arenas. They had two paradoxical moments: first, the rise of environmental consciousness had to be represented in the institutions of the state; second, the contests had to be resolved by political and state means; the paradox was that representation carried the contests from the forests into the heart of the institutions which had to resolve them. In the process, whole fields such as environmental education and law were created, green politics swayed elections, and the bureaucracy evolved to joust in, or judge, the contests. This chapter examines how the claims changed the arenas and what effects this had. The first part deals with the political arena, the second with the bureaucratic arena, and the last part with attempts to resolve the contests during the 1980s and early 1990s.

Environmental politics, industry politics

The rise of environmental values was only one among many factors in the political and state arenas; managing the economy, unemployment, industrial and economic restructuring, and public sector reform all weighed too. Action was spurred or reined in by party advantage, elections, public opinion, the media of the day and conflict occurred within as well as between the sides, political parties, and bureaucracies.

Industry organisations, unions and the environmental movement

The formation of 'sides' in the forest contests masked or suppressed differences within them. On the forest industry side, the national and State forest products associations were formed to counter the environmental movement. They generally advanced the interests of the pulp and woodchip companies and the largest sawmilling companies, rather than the dwindling number of small, independent hardwood sawmilling firms. The Forestry and Forest Products Industries Council

(FAFPIC) and NAFI were so occupied with the proposals for export pulp mills that the problems of small sawmillers, loggers or employees received scant attention. This was most clearly seen in the push for resource security legislation, described later, whose provisions were to be limited to major new projects with investments of $100 million or more, aimed at import replacement or exports. Moreover, the conversion of the old State sawmillers' associations into forest products associations also diminished the attention given to all but the largest sawmillers. This was so much so in Tasmania, for example, that the smaller 'country' sawmillers left in 1991 to form an association of their own.

The forest sector unions, with ACTU coordination, also took an anti-environmentalist, pro-development stance. Their incorporation into FAFPIC facilitated and expressed this clearly; Chris Northover, the secretary of the Pulp and Paper Workers Federation of Australia, was its first chairman. Loggers, mill workers, and truck drivers turned out at demonstrations and meetings organised in support of what was promoted as the industry stance. Their presence enabled the industry bodies to display the fears of unemployment through resource withdrawal, while actual unemployment created by restructuring went unchallenged and the workers got little, except perhaps some better training, for their support.

On the environmental side, differences over ideology and tactics were suspended in alliances for particular campaigns but re-emerged in the face of dilemmas over negotiating claims or incorporation in state processes. The negotiating dilemma arose primarily because matters such as the extinction of species or the exploitation of wilderness have a fundamental, non-divisible and non-reversible character not amenable to the conventional political and industrial modes of negotiating compromises. Even when these difficulties could be overcome, peak groups such as the ACF were in no position to ensure that the disparate environmental groups would agree.

The incorporation dilemma arose as governments, weary of continual conflicts, sought processes in which policies could be set by consensus or negotiation. They invited environmental organisations, or individual environmentalists, to participate with a range of other interests in forums such as Victoria's Land Conservation Council (LCC), Tasmania's Forests and Forest Industry Strategy negotiations, or the national Ecologically Sustainable Development (ESD) process. While such forums provided environmentalists with ideal opportunities to put their case, their participation could be construed as agreement if the final outcome did not accord with their objectives. There were different attitudes and tactics in this dilemma; the more moderate sought participation as the route to influence, the more radical rejected outcomes which allowed logging in native forests. Sometimes participation was continued, sometimes abandoned as a sham, and sometimes decried and left for greater advantage elsewhere.[1]

The swelling cadres of full-time lobbyists employed by both sides became a political factor of their own. Their leadership, and hence their livelihoods, relied on maintaining cohesion among their members, and that needed the continual reinforcement of the threats posed by the other side. To this end, their statements to their constituents were generally expressed in much more extreme language than those to governments or the public. Beyond this, the differences between the two sorts of organisation were marked. The well-heeled industrial organisations could pay high salaries for top executives, hire staff and buy media campaigns; the environmental organisations struggled with inadequate budgets but attracted staff and volunteers to their cause, some dedicating – even risking – their whole lives.

Political parties

Liberal and Country/National Parties

The Little Desert affair in Victoria had shown how the spread of environmental values could cause divisions within the conservative side of politics.[2] Conservation as the wise use of resources had long been supported within Victoria's establishment so that a new Premier, Hamer, was able to win an election in 1972 on an environmentally progressive platform of combating pollution and creating more national parks. However, the progressive views of the Hamer period in Victoria were not the only ones within the Liberal Party. It is, after all, the party of business and capital, and the rise of the New Right ideology of economic rationalism from the mid-1980s led to a further division. The Liberal Party and the Country Party – renamed the National Party in 1982 – usually formed coalitions which contained cleavages between the environmentally progressive policies of the 'wet' Liberals, business interests which wanted cheap access to state resources, traditional rural policies of the Country Party, and the economically-driven, regressive policies of the 'dry' economic rationalists. They all favoured 'States' rights' rather than involve the Commonwealth in environmental matters. In Tasmania, most Liberal backbenchers were so averse to an accord with the environmentalists proposed by their front bench that they even sacrificed a chance to retain government in 1989.[3]

ALP and the unions

The ALP and the unions were also divided on environmental issues. A quarter (27 per cent) of ALP supporters have either joined or considered joining an environmental group, so that environmental issues found ready attention within the party, particularly in its branches and left wing.[4] The party's platforms incorporated detailed environmental policies from 1969, and by the 1980s treated all the major issues comprehensively.[5] However, much of the labour movement was production-oriented and had nothing in common with the environmental movement, which it viewed with a good deal of suspicion.[6] The divide was clearly apparent in the forest sector. Two of the major forest unions were on the right wing of the party and took a strongly anti-environmentalist position, while many of the urban and left-wing branches took an opposite one. Their differences were contested on the floor of the ALP's national conferences. During the 1985 conference, for example, attempts by the party's Environmental Policy Committee to halt woodchipping were successfully opposed by the unions, although, as a compromise, its environmental impacts were made subject to assessment.[7] For the 1989 conference, the unions mounted an unsuccessful case, put together by FICA, that only limited areas in Tasmania should be nominated for World Heritage listing, following the split report of the Helsham inquiry.[8]

The forest unions formed several alliances to advance their views. The Forestry, Forest Products, Printing and Publishing Unions Council was formed in Victoria in 1985 by thirteen unions, including some on the left wing. In Tasmania, a Forest and Pulp Workers Employees Action Group was formed about the same time and although it physically confronted the environmentalists blockading Farmhouse Creek in the Southern Forests, it was thought to be politically ineffective.[9] In 1987 the ACTU assigned a staff member, Bob Richardson, to bolster the efforts of the weak Tasmanian unions, coordinate activities nationally, and represent the union movement on FAFPIC.

Democrats and Greens

While environmentalism was expressed within the major parties and the union movement, it could only win representation in the multi-member electorates of the Australian Senate, Tasmania's House of Assembly and the Australian Capital Territory's Assembly. In the Senate, the Australian Democrats maintained a strong environmental platform and at one stage held eight seats, while the Nuclear Disarmament Party, also with a strong environmental position, secured a seat in 1984 and another in 1987. However, it was not until 1993 when Senators Chamarette and Margetts were elected from The Greens (WA) party, that explicit green representatives entered the Senate. Although the numbers of Independents, Democrats and Greens were small, they could hold the balance of power in that chamber, as when in 1992 the Democrats voted with the Liberals to defeat the ALP government's Resource Security Bill.

In Tasmania, Green representatives were elected to the lower chamber where they even determined which major party formed government.[10] The first Democrat, Norm Sanders, had won a by-election largely on the wilderness and Franklin Dam issues in 1980, and in 1983 the first explicitly Green candidate, Bob Brown, was elected in his place. But it was the Wesley Vale issue in 1989 which proved the turning point when each of the five electorates returned Green Independent candidates – including Christine Milne and Bob Brown.[11] They entered into a fragile parliamentary accord with the ALP which enabled the Liberals to be defeated and the ALP to form a minority government.[12] The accord did not prohibit export pulp mills in general, but specified that one would not be built on the Wesley Vale site. It also specified that more parks and heritage areas would be declared, and the quantity of woodchip exports would not be increased. In the first six months, 1.4 million hectares in the south-west were accepted as World Heritage, of which 1.2 million were made into national park, as was 16 000 hectares in the Douglas–Apsley region on the east coast, a major environmental objective.

But to maintain employment and encourage some development of the State's ailing economy while meeting some of the other environmental objectives was far more difficult. A critical issue was logging in some of the National Estate forests outside the parks. In 1989 the 'Salamanca Agreement' was negotiated between the ALP government, the five Green Independents, industry, unions and the environmental movement, providing an agreed set of conditions for wood production and forest preservation. A Forest and Forest Industries Council of government, industry, environmental, farming and union representatives, with fourteen local working groups, was then set up to negotiate a strategy to show how the accord could be met. The intent, or so the Green Independents thought, was to build the controversial strategy by consensus, but the process was rushed and industry views dominated the council. When it drafted a strategy which included increasing woodchip exports, logging in some wilderness, National Estate and very steep areas, and made no provision to protect rare and endangered species, the environmental representatives pulled out of the process.[13] When the ALP government endorsed the strategy in 1990, the Green

Independents pulled out of the accord. The Labor government and Liberal Opposition then voted together to pass a Bill which enshrined the strategy in legislation.[14] Labor went to the polls in 1992 and was so roundly defeated that, although all the Green Independents were re-elected, the Liberals could govern with an absolute majority.

Elections

Although environmentalists could not win seats in single-member electorates, the prominence of environmental values in the public mind could sway elections, or so politicians believed. The election of the Hawke Labor government in 1983 was held to be tipped by

Barry Unsworth, Premier of New South Wales, unsuccessfully attempted to keep the ALP in government in 1988.
(Geoff Pryor, *Canberra Times*, 29 January 1988)

Wilderness Society poster. Holding governments to the *National Forest Policy Statement* became a key strategy for the environmental movement during 1994. (Andrew Rickerts, The Wilderness Society)

preferences in the marginal electorates of Sydney and Melbourne due to its platform, endorsed by the ACF and TWS, of stopping the Franklin Dam. Its re-election in 1987 was similarly aided when they endorsed it for the House of Representatives, although recommending first preferences to the Democrats for the Senate.[15] On this occasion, their endorsement was due to the ALP's commitment to nominate the forests of Queensland's wet tropics as a World Heritage area in the face of opposition from the State National Party government. The pattern was repeated in the 1990 election when the ALP's environmental platform appeared markedly more attractive than the coalition's, so securing second and third preferences from voters who were attracted first to the many green candidates.

The environmental movement also attempted to influence State elections. In New South Wales, for example, the Wilderness Society ran a 'Save the Forests' campaign against woodchipping in the south-east forests during the 1988 election. It did not stand its own candidates but attempted to get major party candidates in four marginal Sydney electorates to support the cause by publicising the issue and the candidates' stances.[16] In Victoria, similar exercises were held, the prime aim being to get major party candidates to adopt environmentally progressive policies.

The importance of the 'green vote', asserted by the environmentalists and disputed by the industrialists, is uncertain. In any case, it is liable to be overridden by more pressing issues, as the 1993 federal election showed. Then, only about one-fifth (22 per cent) of the people who voted for the Liberal–National Party coalition thought that its environmental platform was closest to their views, and only a little over half (55 per cent) of the people who voted for the ALP thought its policies were closest to their views.[17] An increasing number of people (17 per cent in 1990 and 32 per cent in 1993) felt that there was no real difference in environmental policies between the major parties.[18]

Bureaucratic politics

The paradoxical moments of representation and resolution were reflected in the Commonwealth and State bureaucracies as the new environmental ministries, departments, agencies, units and special programs clashed with the old forest services. New agencies were established and old ones rejigged in attempts to resolve the clashes and the forest contests of which they were an intimate part. All of this was subject not only to the usual inter-departmental struggles for influence, territory, staff and budget, but also to politicking by the forest contestants seeking to protect 'their' agencies or attack those of the other side.

Commonwealth

The Commonwealth set up its principal environment agencies, the Department of Environment and the Australian Heritage Commission, in the 1970s.[19] But, perhaps not foreseeing the coming conflicts, it shut down its small, long-established Forestry and Timber Bureau, absorbing its functions into the Department of Primary Industry.

The forest and mining industries loathed the very existence of the Heritage Commission, and the National Party called for its abolition. Moreover, its statutory independence hardly endeared it to the Department of Environment, on which it was dependent for administrative support.[20] Its staff was slender for the burgeoning nominations to its register (Figure 7.3), but it was helped by its close links with comparable State organisations, professional panels, and voluntary conservation groups, such as the National Trust. It faced all the difficulties of a small, new player in getting access to Ministers, being ignored and having to go to great length to justify its position. For example, in the 1986 dispute over logging National Estate forests in the Eden area, it produced voluminous reports at each stage of the process, whereas the Department of Primary Industry simply asserted that there were no feasible and prudent alternatives, and that delay would result in an immediate loss of jobs.[21] The point is not that misleading statements were made in the heat of battle, but that the larger, older department could assume a mantle of competence, whereas the new agency had to prove itself in the face of continual criticism.

States

The emerging environmental agencies in the States also had to make their way against the forest services and industries; in the process, the bureaucracies became almost as polarised into 'sides' during the 1970s as the contestants. But this could not be allowed to last; the state also had to try and resolve the competing claims with various mixes of amalgamations, transfers of responsibilities, reforms and gradual changes. On top of this, further changes were made as part of the general process of public sector reform being pushed hard in the 1980s. By the 1990s the imperial model of state forestry, established with such difficulty between the 1870s and the 1920s, had been modified considerably (Table 9.1).

Amalgamation

The restructuring started in Victoria, which had already incorporated environmental interests into its Land Conservation Council (LCC) in

Table 9.1 Restructuring of State forest services

New South Wales	
1988	Finances restructured to separate commercial from public goods components.
1991	Forestry Commission adopted a decentralised structure.
1993	Major restructure: • Commercial arm of Forestry Commission named 'State Forests'. • Board of Governance formed to advise Minister on forestry and commercialisation of State Forests. • Forest Policy Advisory Committee formed.
Queensland	
1989	Department of Forestry merged with Department of Primary Industry.
1994	A Queensland forestry corporation to be established within two years.
South Australia	
1979	South Australian Timber Corporation established for joint venture trading and manufacturing projects.
1992	Woods and Forests Department absorbed into Department of Primary Industries and Agriculture and Fisheries – named 'SA Forestry'.
1993	Sawmills, other timber processing, marketing and related functions of SA Timber Corporation and Department of Primary Industries SA Forestry merged into state-owned Forwood Products Pty Ltd. Intended for privatisation in 1995.
Tasmania	
1991	Public Land Use Commission established.
1993	Announcement of intention to replace Forestry Commission with a Forestry Corporation and transfer economic and industrial development responsibilities to Department of State Development and Natural Resources.
Victoria	
1983	Forests Commission absorbed into Department of Conservation, Forests and Lands.
1990	Name change to Department of Conservation and Environment.
1992	Name change to Department of Conservation and Natural Resources.
1993	Plantations placed under Victorian Plantations Corporation.
Western Australia	
1985	Forestry and Wildlife Departments and National Parks Authority amalgamated into Department of Conservation and Land Management.

1970. As the LCC went about its task of reviewing the use of public land, opposition to environmentalism grew. It came not only from industrial groups, such as the Forest Industries Resource Management Group which attacked the LCC's composition, but very publicly from the Forests Commission. The Labor government, elected in 1982 on a strongly environmental platform, confronted the polarisation of the bureaucracies decisively by amalgamating its land management agencies into a single department responsible to a single Minister; the conflicts would be internalised. The forest service, as such, simply ceased to exist, and the forestry–industry nexus of opposition to the environmental movement was smashed.[22] Regional offices were made responsible for all the public forests, parks and reserves, wildlife, fire and land protection operations in their area. Staff of different disciplines and cultures from the old organisations worked in regional teams to integrate the resource management, but the old oppositions were not so readily overcome in the central policy and research units which retained client and disciplinary lines. The upheaval was aggravated by repeated internal restructuring of the department and political changes of its name, first to represent increasing environmentalism, then, when the Kennett Liberal government came to power, a re-emphasis on resource use.

Western Australia followed once the Burke Labor government was elected in 1983, but less drastically. Its amalgamation brought the forest, national parks and wildlife agencies together, but not the lands department, in a Department of Conservation and Land Management.[23] Queensland was more cautious still after the Goss Labor government came to power in 1989, as it integrated its old Department of Forestry into the larger Department of Primary Industry progressively over a number of years.

Transferring environmental responsibility

New South Wales took no steps to stem the antagonisms which had built up during the 1970s, as they had in Victoria. They were deeply felt within the cadres of the Forestry Commission and the NSW National Parks and Wildlife Service, and seem to have persisted – perhaps even hardened – in the 1990s. The commission, which had borne the first onslaught of environmental criticism, was particularly defensive, the environmental movement was forceful, and the State's Land and Environment Court enabled issues to be brought to court.[24] The resulting political and legal battles with environmental organisations and other agencies reduced the Forestry Commission's management autonomy as its operations were subjected to progressively more stringent environmental supervision by other agencies.

The trend started in 1974 when the Forestry Commission was pressed to submit its operations to environmental impact assessment by the State's new Pollution Control Commission.[25] It refused, and instead introduced new policies and a planning system which, it asserted, met the requirements of the State's new *Environmental Planning and*

Assessment Act 1979.[26] There was a revealing difference in emphasis between the *Forests Act* which required the commission 'to take steps it considers necessary and desirable to ensure the preservation of the environment' and the new environmental Act which required it 'to take into account to the fullest extent possible all matters affecting the environment'. Nevertheless, it agreed to prepare EISs for five 'representative' areas to prove its point, though continuing to log them meanwhile.

The environmentalists were quick to challenge the Forestry Commission in the Land and Environment Court over its EIS for two rainforest areas in the north of the State. The judgements found that EISs would be required wherever logging was likely to have a 'significant' effect. But what was significant? and who should determine it? Environmentalists held that virtually any logging was significant, whereas the commission held that most logging was an ordinary part of working the forest. In 1988 the courts ruled that an EIS would normally be required for roading, logging and controlled burning on specific – not merely representative – sites.[27] Applied strictly, the ruling could have stopped almost every logging operation in the State until an EIS had been prepared, assessed and approved for each area. The government first tried to have the commission exempted from the Act but failed to get its legislation through the Upper House. Then it produced a *Forest Strategy* which declared a moratorium on logging in environmentally critical areas until EISs were prepared and assessed. Legislation followed in 1992 which prohibited logging in many old-growth and potential wilderness areas until EIAs or wilderness determinations had been made. However, it allowed logging to continue in fifteen major regions on an interim basis, subject to temporary permits from the Parks and Wildlife Service, while EISs were prepared.[28] Importantly, the Department of Planning was made responsible for assessing their adequacy.

The Forestry Commission complained at the cost and tried to do them quickly, but various inadequacies enabled the environmentalists to get the issues back into the courts. In actions taken by John Corkill and others, the commission mostly lost. The Parliamentary Accounts Committee clearly felt that the commission would have been better advised to accept and improve its EISs rather than suffer such legal costs and embarrassments.[29] The net result of almost two decades of these difficulties in New South Wales was that most of the Forestry Commission's operations were subjected to scrutiny by the Soil Conservation Service and the Department of Planning through the EIA process, and that many could only be conducted with a permit issued by the National Parks and Wildlife Service. The confusion and division of responsibilities caused a hiatus in the State's ability to produce coherent, long-term plans for its forests.

Reform and corporatisation

The wave of reform to Australia's public sector in the late 1980s was economically inescapable; debts and the costs of servicing them had to be reduced, and the public services needed to be 'renewed' as an inherent part of the general restructuring of Australia.[30] The forest services had to be more accountable for the funds they received, objectives and goals had to be explicit, expenditure had to be related to objectives, and any subsidies had to be apparent. Any commercial operations needed to be run in a business-like manner, but just how 'like' was debatable. A New Right conservatism urged the privatisation of government enterprises, a climate of 'economic rationalism' pervaded decision-making in State as well as Commonwealth services, and the corporate model became the vogue for their reorganisation.[31]

The forest services were ripe for reform: they lacked commercial accounting systems which could produce annual profit and loss statements; their accumulated deficits were substantial (Figure 3.3); their economic efficiency seemed poor; and their royalty pricing policies had been roundly criticised as providing concealed subsidies to industry. Yet reform had to recognise how their operations affected the intricate mix of economic, social and environmental values on every patch of the native forests. Only the larger pine plantations seemed straightforward commercial projects.

South Australia and the ACT developed commercial accounting systems for their plantations, followed in 1977 by New South Wales, which commenced a major system for both native forests and plantations.[32] These systems were progressively improved and taken up by other forest services over the next decade, but they still faced severe difficulties in valuing the native forests and in separating the costs attributable to commercial activities from those which provided general benefits to the public, that is between private and public goods.[33] The latter, which included activities such as protection, information services, and the provision of recreation facilities, came to be called 'community service obligations'. The States set targets that commercial operations should make a return of between 4 and 6 per cent on the notional value of the assets employed. When the forest services raised the royalty rates to try and achieve this, the industry campaigned against having to pay for any of the costs of the community service obligations. Greater 'transparency' and national standards in the accounting systems were called for.[34] However, while the accounting systems could separate private and public activities, greater separation could be achieved by restructuring the organisations, or even by radically altering the resource regimes through privatisation.

In 1987, the New Zealand government decided to sell its extensive plantations and state-owned mills, in order to reduce its towering debts.[35] Australian governments watched the move sharply, and there was a tentative proposal to sell some of Queensland's plantations. After it lapsed, governments decided to ape the structure of companies rather than sell their operations ('corporatise' rather than 'privatise'), at least as a first step. South Australia transferred its state-owned sawmills, a joint-venture processing plant and its timber marketing to a state-owned private company, but kept direct control of its plantations. Victoria transferred its plantations to a Plantations Corporation replete with a chief executive officer and board of directors, but only half of the previous workforce.

In New South Wales, the path to corporatisation, strongly pushed by the Liberal–National Party government, was complicated by the need to resolve the inter-agency battles referred to earlier. A new organisational identity, State Forests, was launched in 1993, complete with retitled staff, to undertake the Forestry Commission's commercial activities and general operations in both native forests and plantations. It was to be self-funding and return a dividend to the Treasury. A board of governance was set up like a company board to oversee these commercial activities. Environmental standards were pursued by placing the responsibility for inspection in the hands of the director of the Department of Conservation and Land Management, appointed the Commissioner for Forests for the purpose. Forest policy was placed in the Department of Conservation, and a Forest Policy Advisory Committee was established with environmental, union and industry representatives as well as those from government departments.

Tasmania established a commission for land use planning, hoping to separate the forests to be kept in conservation reserves or national parks from those to be used commercially. It also announced its intention to replace its Forestry Commission with a corporation, and bolster development by placing the economic and development aspects of forest policy in the hands of its Department of Development and Resources.[36] As in New South Wales, the corporation was to be economically driven and the community service obligations were to be separately funded.

Attempts at resolution

Reforms of the bureaucracies was not enough to resolve the forest contests. Although decisions had been made, the political damage when Cabinets could not agree had been formidable and many of the forest issues were likely to re-emerge. Better, more systematic processes were clearly needed for national, regional, and Commonwealth–State decisions. Some built substantial databases to support rational analysis, some developed means for reaching agreement, while others incorporated both aspects, and several were strongly influenced by an industry push for 'resource security'.[37]

National decisions

The Commonwealth, bedevilled by the incessant environmental disputes in several States which culminated in a fifteen-hour Cabinet debate over the World Heritage areas in 1989, needed new processes urgently. First, the Minister for Resources, Senator Peter Cook, tried to initiate a 'forest compact' as a means of reaching agreement between governments, industries, unions and the environmental movement. But he was hardly an impartial broker in the furores over Wesley Vale and Eden's forests at the time, which caused the environment movement to abandon his proposal. Nevertheless, the Commonwealth had other strategies: it improved the factual bases for its decisions; it tried to pass the problem to a seemingly independent body, the Resource Assessment Commission; and it facilitated agreement by at least getting the parties to talk to each other about 'ecologically sustainable development'.

National information systems

The forest contests had been plagued by poor information, as the many inquiries had noted. The State forest services had mapped their forests as timber resources rather than as ecological communities or as repositories of cultural and other values. Moreover, their differing systems made it difficult, if not impossible, to prepare reliable national statements. Standardised national information systems would obviously aid rational debate, but also confer the power to influence it.

The Department of Primary Industries set up a National Forest Inventory to collect mostly mapped data for analysis on geographic information systems (GIS). Ecologists saw the great potential of such systems to provide information on the nation's ecosystems, and urged concepts for its design.[38] But although the Commonwealth's mapping agency had general information, more detailed forest information had to be obtained piecemeal by laborious negotiation with the States. Moreover, much of the initial effort had to go to mapping ownership categories of little interest to ecologists but central to policy debates. The Department of Environment soon set up its Environmental

Resource Information System (ERIN) to collect and make readily accessible many sorts of data, including some of the spatial data collected by the inventory. In addition, the Heritage Commission conducted surveys of wilderness, rainforests and old-growth forests and built databases of the properties recorded on its register. While these systems slowly became able to organise and display selected resource and environmental attributes of the forests, they could not examine policy options for how they might be used.

Resource Assessment Commission

The Resource Assessment Commission (RAC) was modelled on the long-established Industry Commission (formerly Industry Assistance Commission) to provide expert analysis of options for dealing with contentious questions. It was established in 1989 with a core secretariat, and specialist staff for each of the inquiries it was directed to undertake. The first was into the 'options for the use of Australia's forest and timber resources', including those proposed by FAFPIC and the ACF.[39] Its 'policy principles' called for integrating conservation and development, optimising the net benefits to society, and recognising that choices must be made.[40] Its very name construed its inquiries in terms of resources; the environment, defined in anthropocentric terms, was considered as just one of their several values.[41]

The RAC's forest inquiry differed from conventional public inquiries only in the scale on which it was undertaken and the high hopes held for it. It was a mammoth undertaking, with more than 500 submissions, 3500 pages of transcripted evidence taken at public hearings, thirty consultants' reports, substantial analysis and computer modelling by the staff, and weighty draft and final reports.[42] Although it examined mediation, it made no attempt to obtain agreement between the contestants. Its economists tried to assess the 'net benefits to society' of half a dozen options, but their best endeavours to reduce the environmental values of the forests to monetary ones were to little avail; there was no single measure by which the options could be assessed or values traded off against each other. Indeed, the RAC even abandoned the notion of national planning, in its previous form of calculating the area of plantations needed to meet quantity forecasts of future wood demands. Instead, it preferred to let market forces determine it, as of course they will if the shift to corporatisation and privatisation in the forest services takes hold. Most notably, the RAC rejected the environmentalists' call to end all logging in native forests, and addressed its main recommendations to specific issues and restructuring institutions (Table 9.2). Detailed, assiduous, useful and worthy though its reports were, they provided no single, coherent

Table 9.2 RAC Forest and Timber Inquiry, main recommendations

Topic	*Recommendation*
Forest estate	Develop national policies for a permanent forest estate and deforestation.
Old-growth	*Either* stop logging in all stands *or* rank old-growth forests on full range of values, protect adequate examples, and allow logging in remainder, if any.
Wilderness	States should review and protect.
Impacts of logging	Maintenance of forest ecosystem processes should be highest priority for forest managers. Independent audits of logging practices.
Conservation	Fully representative reserve system essential. Active management of forests required to maintain values and benefits.
Plantations	Plantations will develop under normal market forces. No accelerated transfer from native forests to plantations.
Woodchip Exports	Control only as a consequence of land use and forest management decisions. No export tax unless buoyant markets reappear.
Pulpmills	Effluent discharges should be monitored, audited and open to public scrutiny.
Commonwealth institutions	Establish a new portfolio, the Department of Renewable Resources, to integrate industry and environmental responsibilities of forests, fisheries, and possibly agriculture and land management.
Commonwealth–State arrangements	Create a National Forestry (ministerial) Council by including both forestry and environment ministers.
Regional assessment	Develop a national framework, strategies and guidelines for regional forest assessment and planning.

Source: RAC (1992).

national vision for the future of Australia's forests. Rather, the prospects would have to be created as markets provided or destroyed opportunities, region by region.

Ecologically sustainable development

The call by the UN Commission on Environment and Development in 1987 for all governments to plan development in a way that could be sustained led the Commonwealth to set up the Ecologically Sustainable Development (ESD) process in 1990. Instead of the contesting parties submitting their views to yet another inquiry by reputedly wise (usually) men, the ESD process put industrialists, environmentalists, unions, scientists and government officials together in nine working groups covering the different industrial sectors and asked them to find agreed ways for designing a sustainable future. Although the Prime Minister, Bob Hawke, stressed goals of 'the improvement of individual and community well-being' and 'the provision of equity within and between generations', the process was aimed to balance environmental and industrial development issues.[43]

The ESD group on forest use was drawn from forest industries, services and unions, two environmental agencies and two scientists. Although the Australian Conservation Foundation joined in all the other ESD groups, it left the forest group, largely over the issue of logging old-growth. The ACF felt strongly that there was so little undisturbed, ecologically mature forest left that any further logging should be stopped immediately, whereas the ESD group merely recommended comprehensive assessment, examination of the need for protection, and planning to grow some forests to maturity and senescence while others were being logged. Its report foreshadowed many of the RAC's recommendations a few months later.

Consensus, participation and decision

The promise of both the ESD process and the RAC shrivelled in the hands of the Commonwealth for several reasons. Personally, they reflected the consensus style of Bob Hawke as Prime Minister, which changed after Paul Keating displaced him in 1992. Practically, it took so long for their copious recommendations to be sieved and ranked by so many departments and committees that their momentum quite wasted away. Realistically, the ESD process for building agreement directly between contesting parties – less noticeable in the forest group than the others – and the RAC's procedure of detailed analysis, public participation and independent advice exposed the established bureaucracies to outside influences, and earnt the opposition of powerful industrial groups which had expected to do well from them.[44]

Regional decisions

While great principles can be debated at the national scale, it is only at some regional scale that the specific environmental fashioning of the forests can be grasped or realistic prospects for industrial development can be evaluated. Most claims are made regionally, and their social effects are realised in individual lives regionally; thus the contests have to be resolved regionally – as well as at other levels. Two aspects of resolution were pursued by State governments: to aid rational discussion, new data were collected and advanced planning models built to display options for forest use and management; and to aid agreement, processes for public participation were introduced. Here, two examples of the regional wave are described, but there were many more.

Joint Scientific Study of the Eden woodchip area

The contests over the forests in the south-east of New South Wales, which supplied the Eden woodchip mill, were fiercest over three National Estate areas which the environmental movement felt should be kept as conservation reserves. But the larger question of just what areas in the region should be reserved to ensure that the various ecosystems were adequately represented had never been satisfactorily answered; nor for that matter had it been for any region. To ease the political brawling in and between the Commonwealth and State governments, they agreed to a joint scientific study of the question, but allowed it only limited time.[45]

The study team arrayed all the available topographic, geological and climatic information in a computer-based geographic information system to which they added a century's records of plants and animals found in the region, plus much recent survey data. Put together, it enabled biologically sensible 'environmental domains' to be delineated and the representation of each in reserves to be determined. It found that some were over-represented, some under-represented, and proper representation of others could be achieved only with private land.[46] The work was a great step forward in defining the environments, but did not have enough time to develop ecological principles on which new reserves should be designed. It could only report the effects of reserving 10 per cent of each domain (with assessments of 5 and 20 per cent also), but many of the resultant areas had awkward shapes which would have needed considerable adjustment for practical management and ecological stability. There was much else in its report which was harried, mainly by environmentalists, during its preparation and criticised afterwards.[47]

However hurried and harried, the study did provide a quantitative basis from which it would be possible to design a conservation reserve system. But the possibility of doing so was highly politicised that the report was interpreted, misinterpreted, and supplemented by an analysis of 'biophysical naturalness' in a frantic round of negotiations between the Departments of Primary Industry and Environment, and their Ministers. In the end there was a compromise: the industry continued, some reserves were added, and some National Estate areas were to be logged.

Integrated forest management planning in the Otways

Victoria sought to resolve the contests over forest management by combining advances in the factual bases from which rational discussion could proceed with means to reach agreement. Its *Timber Industry*

Strategy had laid down a detailed planning system which required that forest management was to be: economically viable in providing wood, environmentally sensitive, sustainable for all values for future generations, and assisted by public participation in planning.[48]

The Otway Forest Management Area was selected as the pilot area in which regional planning methods would be developed.[49] Intricate geographical information systems and planning models were developed, which enabled the effects of various options for managing the forests to be forecast for 100 years into the future. For each option, the changing age structure of the forest, yields of timber and water, and employment levels were estimated quantitatively. Although the likely abundance of four sorts of wildlife was estimated for every patch of the forest, the techniques at the time did not permit the survival of whole populations to be estimated quantitatively. Hence the conservation of wildlife, like recreation, landscape values and all the other forest functions, had to be planned on the basis of careful assessment, rather than modelling. In all, twenty-eight uses and values of the forests, comprising over 100 individual topics, were covered.

The conventional approach to securing participation from the public at large was followed – by issuing brochures, seeking views, publicising the facts, issuing a proposed plan, and seeking submissions – before making the final one.[50] To this was added a process of securing bureaucratic participation by cultivating links with different specialists and circulating drafts of major documents to the heads of the department's policy branches, each still having historic alignments to particular interest groups. Most effectively, a local Advisory Group was formed, with fourteen members drawn from environmental groups, industries, local government, water boards, four-wheel-drive and bushwalking groups, and universities. All the plan proposals were debated at length in its meetings, which generated many of the ideas which went into the final plan. These structured forms were complemented by extensive discussions with individuals throughout the region and meticulous attention to every point raised. The combination of these methods gave the whole plan credibility across the region. Even if particular local groups felt that their own interests were not paramount in the final plan, the procedure and result was accepted in the region. Not, however, by the Victorian Association of Forest Industries, who claimed it had an 'anti-timber bias', or the ACF, who thought the 'priority seemed to be supplying timber'.[51] The general procedure developed for the Otways was repeated in other forest regions, although with greater difficulty due to reduced budgets and greater complexity in regions such as East Gippsland.

Resource security[52]

The need for State and national lobby groups to push their 'line' even in the face of the most serious attempts to resolve the contests was not confined to the Otways. Even while the RAC was conducting its forest inquiry, NAFI launched a major campaign for new Commonwealth legislation to provide 'resource security' for large investments. It sought legally binding contractual relationships for large investors, and clear, quick negotiating processes. It argued that investment – particularly foreign investment – would never be obtained for large ventures, like export pulp mills, unless the confusion between State and Commonwealth governments was overcome and security was provided against environmental 'interference' to resource rights once granted.

The Primary Industry and Economic Ministers in Cabinet endorsed the notion and the House of Representatives passed the Forest Conservation and Development Bill in 1991. It provided for joint assessment of resource, environmental and social aspects of any region in which industrial developments worth over $100 million were proposed, to be followed by a Commonwealth–State agreement on land use, environmental protection, forest management and other issues. Under the Bill, such an agreement would prevent the Commonwealth from applying its other legislation and powers for environmental protection without providing compensation. At the same time, the Industry and Resource Ministers, Senators John Button and Peter Cook, attempted to curb the power of the Heritage Commission, which they saw as a serious threat to resource security.[53] Although the Minister for the Environment, Ros Kelly, bravely put the case that the proposed regional assessments would enable environmental protection to be planned more thoroughly than ever before, the critics saw the move as an unwarranted abdication of environmental and legal responsibility by the Commonwealth.[54] The environmental movement saw it simply as a means to fast-track pulp mill projects and expose the remaining old-growth forests to development. Moreover, the Bill had little or no support from most of the domestic firms, to whom it offered nothing, and the RAC reported that it was 'largely irrelevant to most of the industry'. The Bill was defeated in the Senate as a result of concerted opposition by the Democrat and Independent members.

The industry was more successful in Tasmania where the legislation to implement the Forests and Forest Industry Strategy, mentioned earlier, altered the regimes appreciably. The sawmillers were protected against the claims of both the environmentalists and the woodchip companies by a State guarantee to make at least 300 000 cubic metres a year of saw and veneer logs available. The pulp and paper and woodchip companies were offered the chance to replace their existing concession arrangements with supply agreements if they wished. But the strategy also increased the area of conservation reserves, deferred others for further study, and set up a system of relatively small reserve areas to complement the 20 per cent of State's forests already in reserves. All told, the area of public forest available for production forestry was reduced from 2.0 million hectares in 1988 to 1.4 million.

Commonwealth–State decisions

After years of politically damaging disputes, the State, Territory and Commonwealth governments finally called a truce in 1992 by signing the *Intergovernmental Agreement on the Environment*; they even included local government.[55] Their agreement called for 'a cooperative national approach' to reduce the number of disputes by sharing data, conducting joint regional assessments of heritage and other values, accrediting State EIS procedures so that one study could serve both masters, and involving the States in international matters, such as the *Convention on Biological Diversity*, and World Heritage listing, in future. How the agreement and the ESD principles might apply became more apparent as a forest policy document was negotiated.

National Forest Policy

The *National Forest Policy* document collated the attitudes and principles which had evolved during the various inquiries, strategies and agreements. It was endorsed by the Commonwealth, States and Territories in 1992 – although Tasmania signed only in 1995, believing that its Forests and Forest Industry Strategy dealt with the issues.[56] The policy set qualitative goals and objectives, and established committees and working parties to straddle the bureaucratic divides, but it did not provide any quantitative targets for such contentious matters as the balances between wood production from native forests and plantations, between pine and eucalypt planting, or between imports and exports. The only specific target was to 'agree that, conditional on satisfactory agreement on criteria ..., comprehensive, adequate and representative reservation system to protect old-growth forest and wilderness values will be in place by the end of 1995.'[57] This was to be realised region by region, but only where the Commonwealth and the State agreed.

In 1994, processes for applying the policy regionally were being worked out.[58] They would apply only if a State nominated a region and invited the Commonwealth to participate in a 'comprehensive regional assessment' which would lead to an intergovernmental agreement. The assessments would go well beyond the joint projects to identify National

Estate values and cover other environmental and resource matters, economic and social issues, and the concerns of indigenous people. The assessments would incorporate the EIS requirements of both governments. It envisaged that the agreements would cover matters such as land use and management, industrial development, forest management planning, dispute resolution, monitoring, and funding. While the proposed comprehensive regional assessments would provide a sensible basis for displaying the options for both preservation and development, and agreements might be arrived at between governments, it is far from clear that the inherent conflicts would be – or perhaps could be – resolved.

Continuing contests

Political parties and institutions of the state are obliged not only to represent conflicting interests but also to resolve their differences. This paradox was not of course confined to the forest contests, but was, as this chapter has shown, sharply apparent in them. Clearly, concern for the forest environment had become part of Australia's public culture in a way which shaped politics and bureaucracies, while mounting industrial claims to the forest resources shaped them too. But the paradox, like the contests themselves, remained largely unresolved.

While scientific research and information systems advanced considerably, the attempts at resolution took a mostly traditional, restricted and particular cast. The traditional processes of inquiries and bureaucracies prevailed, while the few attempts to negotiate State or national agreements directly between the contestants, such as the Salamanca and ESD processes, eventually failed to obtain either environmental or political commitment. Only in some regions, such as the Otways, was some local agreement obtained by informed, unhurried parley. The drive for processes to reach agreement that eventually emerged in the 1990s was restricted to overcoming the obsessive squabbles between Commonwealth and State governments and their bureaucracies, and the federal Cabinet could still become embroiled in heated debates over woodchip exports and old-growth forests in 1995. The attempts at resolution also took a particular cast from the conservative and economic rationalist emphasis on minimising state intervention in the grind of market forces. Lost was any real concern for unemployment, regional dislocation and the social concerns of Australian citizens. Yet, if the best prospects for the future are to be created, ways have to be found to meld social and economic benefits with environmental values. To this, the following chapter is addressed.

CREATING

1990s AND BEYOND

The time has come to break out of past patterns. Attempts to maintain social and ecological stability through old approaches to development and environmental protection will increase instability. Security must be sought through change.

Brundtland Commission (1987), *Our common future*

It will, it seems to me, be necessary for us to ... establish sustainability as the primary objective for enterprises exploiting or using natural resources; to aim for ... a gradual reduction in foreign ownership; ... to devote intelligence and resources to developing our own science and technology and their application; ... and to make access to a healthy, stimulating and dignified lifestyle for all citizens the prime objective of economic policy.

H. C. (Nugget) Coombs (1990), *The return of scarcity: strategies for an economic future*

There is a great sense that new prospects must be created to confront the world-wide environmental and social crises. The Brundtland Commission's report expresses this with its call for sustainable development, as Nugget Coombs does wisely for Australia. The last chapter of this book looks at how the prospects might be created for Australia's forest sector.

Australia seems poised between taking a resolute and intelligent road towards sustainable development in the forest sector and letting current trends lead to social and environmental deterioration. There are promising developments which could make Australia a world leader in creating a new forestry, but Australia is handicapped by an endemic fractiousness between powerful interest groups, which smash long-term hopes in order to make immediate gains. In the end the prospects created for Australia's forests will depend much on Australians' resolve.

CHAPTER TEN

FOREST PROSPECTS IN A CHANGING WORLD

The great sense that new prospects must be created for Australian forests comes not only from present problems, but because 'the future though we canna see, we dread and fear'. With good reason too. Even in Australia, global population pressures, deforestation, resource depletion, loss of species, climate change and raw economic clout may crash cruelly through even the best-laid plans. Experience and resolve can be deployed against them, but experience, with which this book has been concerned so far, may falter if the circumstances are quite new. However, it provides all that we can know of the processes of change, and offers useful categories and questions for prospecting new conditions – warnings, perhaps, rather than sure lessons. There is some circularity in this. The categories with which the history of our experience was written may reappear as those held relevant for the future. This chapter is no exception, as the framework used to analyse the fashioning of Australia's forests in the past is continued to consider their future. Before proceeding, the categories need to be recalled.

Categories

From the instant that a convict's saw bit into the first log at Botany Bay, Australia's forests were inherently part of the changing world order. More perhaps than elsewhere, the landscape was formed by international trade; forests were cleared more for the export of wool, gold, fruit and all the rest than for their consumption at home. Those which remained, with which this book is concerned, were shaped by the mix of international and domestic influences and people – convict, immigrant, born free – from the earliest days. Exports of cedar, then jarrah and karri, and now woodchips, and the replacement of imported timber and paper by domestic production were the most obvious influences. But there were others such as investment, introduced pines and weeds, scientific knowledge, and the swelling influence of conservation and environmental ideas.

The fashioning of the forests was described primarily in terms of the successive resource regimes which assured investment, production, profits for some and work for others. Yet, the forests were never fashioned solely so; some were left to protect catchments, preserve beauty or provide recreation; and rude exploitation became regulated over a century or so as the case for forest conservancy took hold. Then, from the 1970s, the resource regimes were seriously contested by radically different claims for fashioning the forests in an image of nature, in which humans are but one part. The state remained central to all this, which highlighted the importance of the political and bureaucratic categories. The social and regional categories were important too, although often obscured from the public eye by the furore of the environmental contests.

Questions

This chapter looks to the future. The first part questions the highly uncertain forest economy of the Pacific Basin, and the likely changes in the industrial structure within Australia. The next part questions how far the forests may be influenced by environmental claims. These questions are then related to the ability of the present processes of state and social decision-making to deal with the continuing contests. But questioning is not enough; some sort of sustainable future has to be created for Australia's forest sector, just as it has to be globally. Very different time scales apply in this; the Pacific forest economy is so uncertain that considering anything more than a decade or so ahead seems fanciful, yet forest crops take many decades, and ecological maturity centuries. The final part of the chapter addresses the challenge of creating prospects for the future.

Pacific Basin forest economy

The twenty-first century is heralded as the 'Pacific Century' to recognise that the Pacific is economically the most dynamic region of the world, one which may well surpass the Atlantic. Woodchip exports, pulp mill proposals and plantation investments have already shown how Australia's forests are being fashioned – or in the case of imports, relieved – by this economy. Most obviously, more forest products will be demanded in future as populations and consumption continue to increase, particularly in East Asia.[1] Rather than canvass the very uncertain forecasts of demands, the equally uncertain nature of supplies needs to be raised.

Deforestation, depletion and management

The extensive forests of Indonesia, Malaysia, Papua New Guinea and the Philippines provide two-thirds of the plywood, hardwood saw and veneer logs, and half of the sawn hardwood timber traded in the world, as well as meeting domestic demands. The capacity of these forests, and those of other smaller nations in the region, to maintain the 1990s level of supply will certainly dwindle through deforestation, depletion and degradation. The relative extent of the problem can be grasped from estimates that over 60 million hectares were deforested in south-east Asia during the 1980s alone, an area roughly equal to the total area of Australia's forests two centuries ago.[2] Of the remaining Asian forests, many have been logged so crudely that their capacity to support a second cut of sawlogs in a reasonable time is doubtful. Moreover, the regulation of exploitation and assurance of regeneration – so slowly achieved in Australia's history – faces formidable difficulties due to the desperate need for export income, endemic corruption in some countries, illegal logging, and the scientific problems of developing successful silvicultural practices in the highly diverse forests of the wet tropical lowlands.[3] Nevertheless, better management has been encouraged since the 1980s by strenuous efforts under FAO's Tropical Forest Action Plan and by the International Tropical Timber Organisation (ITTO). Some reductions in logging and exports have occurred voluntarily, and various community forestry schemes endeavour to meet local needs.[4] The major regional trade question is whether production will be reduced to some relatively stable level that can be maintained in the long term, or whether more forests will be mined to virtual exhaustion, as in the Philippines and Thailand.

Preservation and environmental management

The world-wide drive of environmentalism to dedicate parks and conservation reserves, and the struggle of indigenous forest-dwelling peoples to preserve their domains may reduce timber supplies from tropical countries somewhat, but the major effect is already occurring in the great conifer forests of the Pacific north-west coast of the United States and Canada. These forests supply more than half the softwood sawlogs, sawn timber and pulp traded internationally, but their remaining old-growth stands have become highly valued for the habitat that they provide. As in Victoria, those in the United States have to be managed to sustain the full range of uses and values of the forests, and to preserve rare and endangered species. When the federal forest service was unable to show that its plans could do so for the endangered Northern Spotted Owl, a court suspended logging in several of federal forests finding that:

> The argument that the mightiest economy on Earth cannot afford to preserve old-growth forests for a short time ... is not convincing today. It would be even less so a year or a century from now.[5]

The subsequent revision of forest management to ensure the sustainability of the populations, not only of owls but of some 1500 species, led to a decline in the volume of sawn timber being exported. As in Australia, it is uncertain how much production throughout the Pacific will be reduced by the creation of additional conservation reserves and by managing the remaining production forests on much longer rotations than have been envisaged in the past.

Plantations

Another uncertainty is the extent to which new industrial plantations will be able to offset shrinking supplies from both hardwood and softwood native forests in the Pacific Basin, while at the same time meeting rising Asian demands.[6] Brazil, Chile, New Zealand and Venezuela have substantial plantations of fast-growing exotic species, and will soon be joined by Fiji, Thailand, Indonesia and other countries of the region. With intensive management, tree breeding, ground cultivation, chemical weed control and fertilisation in what is forestry's Green Revolution, these plantations will produce large quantities of uniform raw material, largely for the pulp and paper industry. For example, Oji is establishing plantations in Thailand as well as in Australia; a Finnish company, Enso–Gunzeit, has planned a joint venture to create a 100 000-hectare plantation in Kalimantan to support a chemical pulp mill producing 500 000 tonnes a year in ten years' time, and a joint Indonesian-Japanese venture in Sumatra will draw on 140 000 hectares of plantations to supply a pulp mill expected to begin exporting to Europe and China in 1997.[7] It is uncertain how far these new, fast-growing plantations will compete with poorer-quality resources from old-growth mixed hardwood forests, such as those supporting the Australian woodchip export trade.

Partner, servant or victim?

It is inconceivable that the highly dynamic economy of the Pacific Basin will not affect the forests on its Australian periphery; it is only the extent, timing and role for Australians in their continuing import and export of forest products that is in question. Although impending timber shortages have long been forecast and an eventual world timber famine can be posited with a certain Malthusian logic, the large stock of standing timber and the option of growing at least pulpwood quickly means that it is not imminent. Rather, Australia can expect to remain in a competitive market for the next decade or two at least.

The price competition from imports will doubtless fluctuate, encouraging or discouraging domestic mills and plantations in the process, but it is the chance of exports that might have the greatest effects on the forests. The roles taken and their consequences may well vary from region to region. At best, Australian firms can be full 'partners', in either trade or joint ventures with Asian companies. These are becoming more common because they provide mutual access to resources and markets. They also have disadvantages if the exporting partner wishes to increase investment and upgrade production, say from pulp to paper, and the importing partner does not.

A less favourable role is that of 'servant', in which Australian firms are contract suppliers of raw materials for off-shore processing. It provides limited employment and devotes local resources and some capital primarily to the benefit of the importing company. This is the role for most of the woodchip industry at present.

The least favourable role is that of 'victim'. The ever-increasing dominance of multi-national corporations is leading to production in some sectors being relocated in regions of 'comparative disadvantage' in the world economy, where governments have little choice but to ensure pliant labour, accept more pollution, or open their resources for exploitation cheaply. Many developing countries are in this situation, exacerbated by long-term declines in prices for their exports of primary products and resources. Parts of the Australian forest sector may take this role. Initially, the woodchip industry did not pay enough in royalties to cover all the public costs. Now, hard-pressed farmers drawn into growing crops of eucalypts for the woodchip export trade could find themselves in difficulties if the price deteriorates. Working communities dependent on such single export commodities are particularly vulnerable.

Australian industry

The endless restructuring of industry in the economic 'war of all against all' involves categories of firms small and large, local, national and multi-national. Each of the trends in the continual centralisation of production and the related concentration of ownership raises questions for the future.

Centralisation and concentration versus persistence and devolution

The trend to centralise production in fewer, larger mills in fewer hands will doubtless continue, but has obvious limits once an oligopoly position has been reached. There is little room left for further concentration in the pulp and paper industry, for example; the bulk of hardwood sawn timber production has passed from small, local, family-owned mills to much larger company mills; while pine sawmilling and panel manufacturing are clearly in the hands of large businesses.[8] The resource regimes and consequent forest fashioning will continue to be driven largely by the increasing size of the large mills, leaving the place of the small mills in question.

The most important of the small, local businesses are those which draw on small quantities of premium quality logs and process them into furniture or similar products for niche markets. They are vulnerable to losing their special, long-term resources in the sweep of large-scale logging across the landscapes, and the adoption of short rotations for the pulp and woodchip mills. Two other sorts might be able to survive at the periphery of the sector. First are the remarkably persistent small sawmills which draw on small parcels of logs and supply local rural markets. Second are the few mills in each region which by ingenuity and hard work manage to saw palings, pallets and case timber from logs that others would reject or use for pulpwood. These will be able to survive – be encouraged even – only if the regimes are administered appropriately.

Another group of small businesses is those which supply materials or services to the large firms. The general trend of large firms in the 1980s and 1990s to devolve some of their operations to contractors will continue in the sector. It is, of course, almost a defining feature of logging and can be expected to become more common in forestry as the trend to corporatisation of forest agencies and privatisation of plantations continues.

Import versus export of foreign investment

Concern with foreign investment has been largely directed either to the benefits of attracting capital inflow, or to its dubious influence on the domestic industry and economy. However, capital has no allegiance to country: if better or more strategic investments are available elsewhere, Australian or any other capital will go there. In the forest sector, both Boral and Amcor have actively expanded overseas. Amcor invested $1.8 billion in two years to 1994 in paper and packaging mills in Britain, France, Germany, China, Malaysia and the United States, and operates in fourteen countries.[9] The point is that Australia will not necessarily be chosen for future investment by Australian-owned companies if there is greater advantage elsewhere.

Looking more widely in the Pacific region, the web of corporate interests is becoming increasingly internationalised in ways which connect with Australia. For example, New Zealand companies have invested in forests, plantations, mills and merchants in Australia, Brazil, Canada and Chile. And when its state plantations and sawmills were privatised, New Zealand firms obtained only 40 per cent, the rest being sold to investors and joint ventures from China, Japan, Hong Kong, Malaysia and Singapore.[10] Large American pulp and paper companies typically have converting, packaging or other operations in half a dozen other countries; Kimberley–Clark for example, which jointly owns the tissue mill in South Australia with Amcor, has an expanding world-wide network to produce its Kleenex brands. But networks can shrink or be sold easily enough, giving 'host' countries unexpected 'visitors', as Australia found when the UK-based Bowaters sold its tissue mill at Myrtleford and other Australian operations to New Zealand's Carter Holt Harvey in 1994.[11] Japanese paper companies with Australian connections – like Daishowa, Daio, Itochu and Oji – are investing in plantations and mills elsewhere in the region and joint-venture mills in China.[12] Some corporations which have not previously been involved in the forest sector, like Shell and British Petroleum, are also starting to invest. And there are many other international linkages through consulting engineers, management and marketing arrangements. Looking to the future, it seems inevitable that Australian industries will become increasingly enmeshed in the Pacific web of capital. Thus, the future industrial structure appears as one dominated by large companies or corporations, intimately linked with international interests, and surrounded by niche, peripheral, and dependent small businesses. The extent to which the forest industry remains in any real sense Australian is likely to be influenced most by the paucity and dispersion of the forests, and the small size of the domestic market.

Existence, ecology and forest regimes

The environmental challenge to the industrial fashioning of the forests was far from resolved during the two decades of contests described in the previous three chapters. Certainly, public consciousness was raised, environmental concerns institutionalised, areas of parks and reserves increased, logging practices improved and science advanced, but the environmentalists' audacious claim that logging should cease in all the native forests was rejected by public inquiries and seems never likely to be achieved. However, the goal of having a system of representative conservation reserves was accepted, though not fully achieved. Even if it had been, it was realised that it could not alone preserve Australia's biodiversity if the surrounding landscapes were so battered. By the 1990s, it had become clear that both paths will be needed for the future: continuing to demarcate conservation reserves and wilderness areas until an adequate system is assured, while managing the public and private forests outside this to conserve a range of values and uses beyond commodity production. In either case, the forest contests will continue, because there is unlikely to be agreement on the adequacy of any reserve system, and changing the forest management would alter the flows of wood – most markedly from the woodchip export concessions.

Even if the contests between the established industry and environmental claims could be resolved, new difficulties will arise because the pressure for exploitation will increase as the forest economy of the Pacific expands. The pressure for environmental protection will also increase in response to this, as a result of climate and other changes, and as ecologists discover more about the intricate life of the forest.

Exploiter, preserver or conserver?

The question is what role will Australia choose – or be economically free to take – in this difficult future? The 'precautionary principle' might be invoked to favour preservation until the environmental consequences of developments are known, but a similar 'principle' could as easily be invoked to secure development unless irreversible environmental damage was certain. Here, the choice is posed as being between 'exploiter', 'preserver' and 'conserver' roles.

Exploiter

The simple role of forest exploiter – taking resources without renewing or paying fully for them – is natural to individual capitalists, who must profit where they can or fail. It characterised much of the general industrial history recounted in the first chapters of this book, for even when forest conservancy gained hold, exploitation was only made more orderly. The sawlog resources of the native forests continued to be run down, even during the 1980s, and endeavours to counter this drained the public purse from the 1950s. Even in the 1990s it is far from clear that industry pays its way.

The state faces the dilemma that it must concede resources for development and production, yet do so in ways which renew the

Regrowth of spotted gum aged twenty years, Benandarah Heritage Area, NSW, which has been managed as a State Forest for over a hundred years. (State Forests, NSW)

resources and yield public benefits greater than the costs. Its resolve to avoid the role of exploiter can too easily weaken when the benefits trickling from exploitation occur in the present, while the greater benefits of forbearance occur in the distant future, as they do with forest crops. Moreover, as forest service budgets tighten, less care can be taken to regenerate and tend the cut-over forests. Worrying signs that Australia may be slipping back to the role of 'orderly exploiter' emerged in the late 1980s and early 1990s. Ironically, the drive to corporatise, rationalise and streamline the forest management agencies reduced their staffing (Table 5.1) at the very time that the exponents of fine principles, policies and strategies were calling for the forests to be managed for a wider range of values than ever before.

Preserver

The simple role of preserver was established later, but strongly in Australia at several levels. The representative reserve system proposed in 1974 and accepted as a goal in the 1990s will doubtless be slowly, if not fully established – it may be recollected that it took more than forty years for the target area of state forest to be reached (Figure 3.1).[13] Scientific arguments strongly supported the preservation of wilderness, rainforest and old-growth forests on the grounds that their particularly rich biodiversity warranted their reservation first. But it was spiritual and popular cultural icons of these forests, koalas, certain rare possums and potoroos, which weighed substantially in the public mind and political fray. Further types of forest or particular creatures may well be so elevated in future. An ecocentric philosophy may be necessary for spiritual health, but arguments for a more frugal lifestyle and a spiritual revival seem unlikely to upset Australia's materialistic ethos. Nevertheless, the sacredness of nature that they proclaim may add political weight to preservation, irrespective of alternatives.[14]

Conserver

The conservation alternative is the hardest role, difficult to learn, more still to perform. Yet it is the essence of the human enterprise; we must fashion nature to survive. We can strive for ecologically sound ways, and must do so wisely to survive long. This necessarily involves recognising and balancing multiple objectives, and appraising all the uncertainties of both environment and development.

The concept of conservation as wise use drove the rise of forestry from the nineteenth century. It has been too easily claimed, under the banner of 'sustained yield, multiple use', to justify much of the management of Australia's forests. It needs to be restored to its full meaning, one which extends beyond the long-term regulation of timber yields to embrace all the functions and beauty of the forest, and to sustain the working communities dependent on it for their livelihood (Table 0.2). It is one of the major strands in classical forestry, and its essence was captured by Köstler when he concluded his text on silviculture with a quotation from Goethe:

> Our whole attention must be directed to observing carefully the processes of Nature, so that we do not make her refractory by preremptory orders, but,

> on the other hand, do not allow ourselves to be diverted from our purpose by her caprice.[15]

Internationally, the banner of multiple-use, sustained-yield management is being dropped, and the concept of forest conservation is being restored and enlarged under the banner of 'ecosystem management' or 'forest health' as:

> attempts to maintain the complex processes, pathways and interdependences of forest ecosystems and keep them functioning well over long periods of time, in order to provide resilience to short-term stress and adaptation to long-term change. Thus, the condition of the forest landscape is the dominant focus, and the sustained yield of products and services is provided within this context. ... ecosystem management must maintain our social and political systems as well as meet our need for both consumptive and non-consumptive uses of the forest.[16]

Philosophically, ecosystem management implies a shift to a more ecocentric attitude; scientifically, it requires a shift from the silvicultural focus on individual stands of trees to one on the forest landscape as a whole. Its recognition in Australia would require a radical reappraisal of how the forests are managed, and that would be a most contentious matter.

The case of the concession for Harris–Daishowa's woodchip mill at Eden illustrates the severity of the potential contests. The initial plan was to clear-fell the entire forest over forty years. Although a 'chequerboard' pattern of cutting was introduced, and nest trees and strips were retained, the forest was to be one of perpetual youth. By contrast, a plan prepared on the emerging ecosystem management principles of the 1990s to maintain the full range of environmental values might well let the forest grow much older and envisage felling over, say, 160 years. However, the mill has already been built and half the forest cut to the initial plan. With the benefit of hindsight, it can be seen the forest has been severely overcut, but a transition to a conservation basis would be difficult to achieve because it would drastically reduce the quantities being cut and exported, with consequent effects on people's livelihoods and the regional economy. The situation was exacerbated because the mill allocations were not reduced when the resource was diminished by the reservation of many small areas and some large parks. Although Eden and the Victorian ash forests may be extreme cases, the need to lengthen rotations in the native forests from industrial to conservation principles is a general one which will present widespread difficulties.

The future establishment of plantations is related to the management of the native forests in three ways. Plantations could provide additional pulpwood resources fairly quickly, as part of strategies to make the transition to conservation principles in the native forests socially and politically feasible. They could provide additional resources when the limit of the productive capacity of the native forests, properly managed, has reached its limit. Also, their wood can replace wood from the native forests in many, but not all, uses. However, their development too needs to be placed on a sustainable basis. This raises still more questions about their environmental impacts, particularly when they are intensively managed on short rotations with high levels of chemical and energy inputs.

People

The discussion of the forest contests in this book cast the industrial and environmental organisations as the leading actors, while labour was depicted as the subject of market forces, only occasionally striking against the employers or protesting against the environmentalists. Similarly, regional and social benefits were depicted primarily as the consequences of industrialisation, restructuring or resource reduction, variously influenced by state policies for resource allocation and decentralisation. Neither workers nor the people generally were seen as leading actors with much direct – rather than representative – influence; the initiative stayed with capital, latterly challenged by environmentalism and by the desire for public participation in forest decisions. The question at issue for the future is whether the social benefits of the forests flow to, or are subverted from, the people; firstly, in relation to society at large, and secondly, in relation to employment and the diffuse benefits from production.

Beneficiaries or subsidisers?

The first question asks whether the total benefit to Australia from the industrial use of the forests will be greater or less than the costs. Of course, many of the functions of the forests, such as catchment protection, beauty, carbon sequestration and oxygen emission and the provision of wildlife habitat (Table 0.2), confer great public benefits; in economists' words they are public goods, not privately appropriated or traded, whose value may be diminished – rarely enhanced – by the operation of the resource regimes. Both the preserver and conserver roles seek to maintain them. But the resource regimes, as discussed throughout this book, allocated public resources to private ends. The public fared economically according to how the forest services

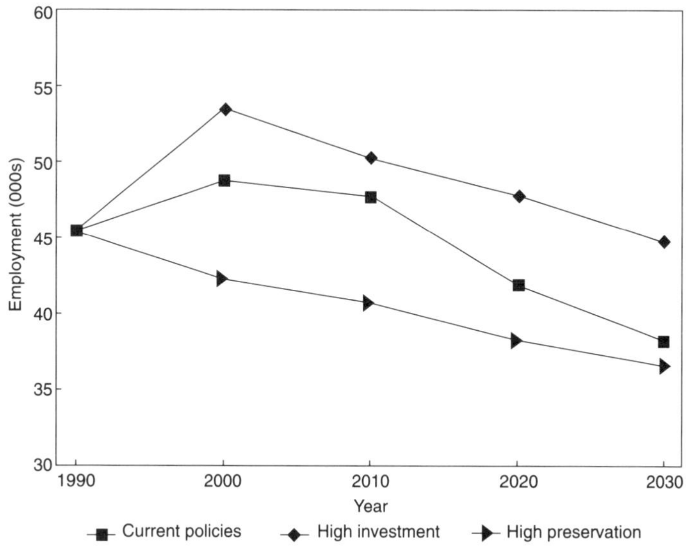

Figure 10.1
Employment forecasts for forest sector, 1990–2030
Source: RAC (1992), vol. 2a, p. L92.

managed the resources and their budgets. They have not fared too well so far, because the timber resources of the native forests – the public's assets – have been grossly run down, with little net income to the Treasuries and substantial debts for the plantations.

Optimistically, one might conclude that the era of economic exploitation has ended: the assets are virtually as low as they can get and current plans will see them slowly increase; royalty rates have been increased to more economic levels; the economic rationalists' arguments will lead to greater efficiency; their push to privatise the plantations will see an end to subsidisation. Alternatively, one might fear that the plans will fail under pressure to cut crops early for woodchip exports, for example, or to drop royalties below the costs of regenerating and tending the new crops.

Workers and unemployed spectators or working communities?

The second question asks whether the industrial use of the forests will provide employment for more or fewer Australians. This is important because employment, or its lack, is the prime indicator of social welfare in the 1990s and is closely correlated through wages to regional prosperity. It can be increased by expanding production or by further processing the wood to higher-value products, and it can be reduced by mechanisation, automation and restructuring.

The Resource Assessment Commission estimated future employment according to several scenarios, three of which are displayed in Figure 10.1. Its forecast of current policies seems over-optimistic, in view of more recent trends (Table 5.1). Even the additional employment created by its most expansionary scenario (based on FAFPIC's *Growth Plan*) would be negated eventually, while the most preservation-oriented scenario (based on ACF's policies) shows a continual decline. The Industry Commission was asked to examine the potential for further processing, and hence for employing more people, but believed it would be most efficient to leave such things in the hands of uncurbed market forces.[17]

Such reports are truly 'lifeless' because they treat people's livelihoods as the consequences of economic forces rather than their purpose. Of course, it is true that economic forces are important, that many are external to Australia, and that failure to keep industries competitive can end in disaster; but other, more strategic approaches are possible. Nationally, there are proposals to reorient economic policy from the economic fundamentalism reflected in these reports, towards restoring full employment through more public funding of community services and infrastructure, and by other measures.[18] To make people's livelihoods the purpose of economic policy would require such reports to be recast towards devising policies for working communities. This is not to advocate subsidies or allocation stratagems to protect inefficient industries; rather, it is to ensure that the structural changes which reduce employment are intimately linked with those that increase it. For example, sawmillers wanting to amalgamate log quotas from smaller mills they purchase might be made responsible for maintaining employment by installing further processing equipment, developing new products as the Black Forest Sawmills did, or developing new markets. Such activities could well be assisted to start, but should then be economically viable in their own right.

Strategies to assist working communities in the long term are compatible with the conserver role, but the preserver or exploiter roles lead to their collapse. With the benefit of hindsight, it can be seen that the overcutting of the Eden area, for example, might have been averted. A much more enduring future could have been created for an admittedly smaller working community if planning had been both socially and environmentally more far-sighted. Transitional strategies will be needed for this and other working communities to enable them to move to a conserver role with a long-term future.

State

The state in its several levels will remain the arena, judge and marshal of the forest contests, with some of its various institutions entering the lists as champions of powerful interests and noble causes. But its freedom to create future prospects is bounded by the international economic forces discussed earlier, and world environmental politics which must be reckoned with.

Commanded or commanding?

By the 1990s the international influences on conservation and preservation had become 'global' through a number of international environmental agreements and conventions relevant to the forests (Table 10.1). The extent to which they will be honoured is uncertain, as little more than disparaging comments apply to breaches. Nevertheless, Australian governments do want to be seen as at least trying to implement their obligations. They are no foreign imposition, but a synchronous expression of world-wide environmental concern in which Australia takes an intimate, sometimes leading, part. For example, the international *Convention on Biological Diversity* in 1992 was preceded in Australia by submissions to the Prime Minister in 1989, the start of the Ecologically Sustainable Development process in the following year, and the formation of a Biodiversity Unit by the Commonwealth government in 1991. The convention was ratified by Australia in 1993 and a National Biodiversity Strategy was prepared for agreement by Commonwealth and State governments in 1994. It would seem likely that the influence of international agreements will strengthen in future. Pressure may also be exerted through markets as consumers are urged to buy 'environmentally friendly' products. This has already influenced the trend away from chlorine bleaching in the pulp and paper industry, and might influence forest management if the moves for certification of timber as coming from sustainably managed forests take hold.

International agreements represent a passage of some national power to the international arena, and may lead to a consequential passage of some State powers to the Commonwealth, as the Franklin Dam and Queensland rainforests cases showed. To environmentalists, this has been a means of bringing enlightened attitudes to bear, but to industrialists, State politicians and their officials it has been an unwarranted intrusion on their right to use the forests as they will.

It is uncertain how far such influences will withstand economic commands, and how they will be articulated between Commonwealth and State governments. Optimistically, one might look to UNCED's world conference at Rio de Janeiro and in Australia to the proposals for Commonwealth–State regional agreements as examples of the willingness of governments to work cooperatively together. But such truces are liable to be broken where contesting interests or

Table 10.1 International instruments possibly relevant to forests

Year	*Instrument*
1971	Convention on wetlands of international importance.
1972	Convention concerning the protection of world cultural and natural heritage.
1973	Convention on international trade in endangered species of wild fauna and flora.
1974	Australia–Japan: Agreement for the protection of migratory birds.
1976	South Pacific: Convention on conservation of nature.
1979	Convention on conservation of migratory species of wild animals.
1983	International tropical timber agreement.
1986	Australia–China: Agreement for the protection of migratory birds.
1986	South Pacific: Convention for the protection of the natural resources and environment of the South Pacific.
1992	Arising from the UNCED Conference: • Convention on biological diversity • Framework convention on climate change • Statement of forest principles • Agenda 21 • Rio declaration on environment and development.

governments see advantage.

Against this relentless 'globalisation' of affairs is set the regional nature of forest biodiversity, resource management, industrial development, and their realisation in the livelihoods of working communities. Yet this level is virtually unrepresented in the affairs of state. Local government has no constitutional command of forests and generally covers smaller areas than forest regions. Optimistically, one might look to the formation of regional planning, coordination or advisory bodies to provide regional expression. Many have been useful for facilitating state planning at the regional level but, lacking consititutional power to raise revenues or to legislate, they have been ineffective voices for the people. For the future, they are clearly needed to give a voice to working communities, to examine the biology and resources of their specific forests, and to provide a level at which contesting parties can meet face to face. However well they function, they are vulnerable to being over-ridden by State commands.

Champions, judges or peace-makers?

The state's various institutions and processes, described throughout this book, have reflected the ideas and contests which have formed and shaped them. Various strands will need strengthening to deal with changing circumstances and continuing contests in the future.

One strand is to continue the advance of forest science and rationally arrayed resource and environmental information. This is clearly necessary both for applying ecosystem management principles to the forests, and for protecting them from new hazards that may arise. Equally, technology for forest products should continue to be developed in relation to pollution abatement and new products and processes to add greater value to the wood.

One strand is pitifully weak in Australia: the monitoring of the rapid changes in the Pacific Basin forest economy. It needs considerable strengthening so that Australian industrial and trade policy is alert to the changes and can be revised in a strategic manner. It is important that information be kept up-to-date, not only in the conventional terms of resources and trade, but also in terms of industrial ownership.

Perhaps the most critical strand concerns the social and political processes for deciding how the forests should be fashioned. It raises thorny questions beyond the scope of this book, concerning the nature of representative and participative democracy, and the rights of unborn generations and non-human species. But pragmatically some 'better' ways to resolve the continuing forest contests are widely thought to be necessary. There is probably little agreement on what might constitute 'better' – would it be quicker or more thorough? However, it is clear that it is necessary to build respect between the contestants and the state agencies as a first step. Optimistically, one can look at the 'open government' approach of making all the scientific and resource information readily available. The ESD, Salamanca and Otway cases provide examples of processes which might build agreement from informed discussion and respect. But these were demonstrably fragile and showed the task to be formidable.

Roads and resolve

This chapter has assumed that the alternative paths suggested in the categories of economy, industry, environment, people and the state are those between which we have to choose as best we can. But such a pragmatic, reformist, parochial outlook is held by some grand theorists to be so inadequate as to be part of the problem. Envisaging a world headed for environmental collapse, they see nothing less than 'zero population growth in the lifetime of our children',[19] 'a cultural renaissance',[20] or 'restructuring the global economy, ... major shifts in human reproductive behaviour and dramatic changes in values and lifestyles'[21] as being essential and urgent. They proffer principles, discussions, and exchanges of information between scientists and

Alpine ash trees resulting from group selection logging following clear-felling in 1916, Bago State Forest, NSW.
(State Forests, NSW)

'concerned and interested persons', but no strategies to accomplish such revolutionary changes in the face of powerful interests.

This does not seem an age for revolution. The fate of the Russian one – which is the only one in this century to have attempted change on a scale comparable to that envisaged by the grand theorists – casts doubt on the feasibility of great revolutionary change, and shows the inherent weakness of central planning to handle social complexity.[22] By contrast, capitalism and liberal democracy are held to have great, perhaps unique, capacities to handle complexity and to adapt to changing circumstances. Moreover, democracy by providing the right to participate in the running of the state fulfils deep human desires for social recognition. Although Fukuyama argues that the combination of capitalism and liberal democracy constitutes the final, universal and homogeneous form of the state,[23] the contradictions inherent in capitalism, economic and population growth, and all their environmental problems cannot be dismissed. New roads to the future are needed, as the Brundtland Commission found, but their construction has to be started by the institutions we have and has to deal with the complexity of the present and the uncertainty of the future. They can only be constructed place by place, year by year, sector by sector, category by category, doubtless with endless hard contests for livelihood and democracy, conservation and utility, beauty and species existence on the way.

Low road or high road?

The concept of 'sustainable development' advanced by the Brundtland Commission conveniently herds together a number of issues.[24] Its very vagueness and inherent contradictions capture the complexity and uncertainty of creating new prospects for the future.[25] Even if it cannot provide an unambiguous objective, it does provide criteria against which many possible roads forward can be assessed and the obviously unsustainable discarded; in this light it is better seen as a journey than as a destination.

Criteria for assessing the forest sector, derived from the previous discussion, are shown in Table 10.2. Any assessment would be contested, and there is no aggregate measure that would enable progress in one category to be traded off against another. The nature of the social and political choices can be highlighted by considering two of the many possible roads forward. The 'low road' is a pessimistic one which continues present trends without any energetic or intelligent measures to deflect them. Conversely, the 'high road' is an optimistic one which envisages Australian society taking more deliberate steps.

Low road

Much of the fashioning of Australia's forests during the 1980s and early 1990s proceeded patchily and partially along the low road towards sustainable development. Although economic efficiency improved and subsidisation decreased, foreign ownership also increased and employment fell drastically. The maintenance of a wide range of enterprises, large and small, corporate and family, established and

Table 10.2 Criteria for assessing sustainable development

Category	*Criteria*	*Measure*
Economy	Efficiency Flexibility Role	Competitive? Adjustable to changing circumstances? Partner, servant, victim?
Industry	Ownership Size Small business	Australian or Foreign ? Open to various size firms? Independent, peripheral, dependent?
Environment	Role	Exploiter, preserver, conserver?
People	Citizens Workers	Beneficiaries or subsidisers? Creation or destruction of employment?
State	Australian–International Commonwealth–State State–Local Bureaucracies Process	Participant or isolationist? Co-operative or hierarchical? Co-operative or hierarchical? Integrative or sectional? Representative or participatory?

entrepreneurial, would have been desirable strategically and socially, but governments have been either lazy, or so enchained by the arguments of economic rationalists, that the trend to concentration and centralisation was left to run unheeded. Similarly, the Commonwealth virtually abrogated its control over the extent of foreign ownership. Although more conservation and wilderness areas were reserved, and immediate logging impacts were ameliorated, conservation in its wider sense of sustaining all the values of the forest was not fully recognised. Most seriously, the future of working communities was hardly recognised at all. Although a great paper mountain of worthy reports, noble policies and fine strategies has accreted, the real fashioning of the forests was little changed. Indeed, the limited advances that have been made could easily be lost in the conservative agenda of corporatisation and reduction of the public sector; if there are not enough foresters, supervisors and workers, the forests can never be protected and tended carefully.

The forest sector would find it hard to climb out of the low road. The most severe resource problem is to find transitional strategies in situations where working communities and established industries have been built up on plans which over-committed forest resources and which are now seen to be unsustainable. Historically, forest services were unwilling or unable to deal with such problems until the resource became so depleted that collapse was imminent. We can have little confidence that this era has passed; even the pine plantations in the Australian Capital Territory – the easiest of all resources to plan – were overcut to the extent that quota reductions were forced in the 1990s. Moreover, the general need to lengthen rotations to accommodate values other than wood production will greatly increase the difficulty of the problem throughout all the native forests.

Overcoming these and other problems will be extremely difficult in the face of the bitter hostilities that have accumulated between the contestants and between some of the forest and environmental agencies. Their contests have been primarily over territory, that is between the preserver and exploiter roles. It will be very difficult to escape from this polarisation and admit into the debate conservation, the future of working communities, and all the real complexity implied in 'sustainable development'.

The constitutional structure and political process also make it difficult to take the long-sighted views necessary to climb out of the low road. States which are hungry for development can be played off against each other by developers seeking concessions; States may take on expenses knowing that the Commonwealth has ultimate fiscal responsibility; politicians with an eye to the next election may make symbolic declarations rather than pursue substantive actions, and the short cycle of elections directs their attention away from long-term issues. These matters apply widely across environmental policies but have particular force for forests with their long timetables.[26] Moreover, it will be difficult to escape from the State–Commonwealth antagonisms; for the forests these reflect the upwards transfer of power as much as the environmental–industrial contests themselves.

A further paradox occurs between the increasingly centralised, even globalised, future and the necessity of dealing with the issues of environment and development in detail, region by region. Although regional planning processes are being developed, their main focus is on overcoming the State–Commonwealth antagonisms. It will be difficult for local communities, initiatives or governments to gain a place in these processes, or to assure a future for their working communities.

A revealing episode

The difficulties of climbing out of the low road were laid bare during the summer of 1994–95 when the woodchip export licences needed to be renewed. Although they were long-term licences for ten to fifteen years, they had to be renewed annually in light of advice from the Commonwealth's Environment Minister, but it was left to the Resources Minister to decide or take the issue to Cabinet. At the end of 1993, a few coupes intended for cutting in 1994 were withheld and, with the

National Forest Policy Statement just signed, it seemed that logging would be banned in old-growth, biologically rich forests, at least until they had been assessed. But by the end of 1994 the situation had changed: there was a new Environment Minister, Senator John Faulkner; the environmental movement was disillusioned with the way the statement was being ignored; and the industry was pressing the Resources Minister, David Bedall, for expansion.

The new Environment Minister took up the issue vigorously and gave environmental organisations $100 000 to find the biologically rich areas at immediate risk. Their answer may have been meticulously prepared, but could hardly be seen to be as impartial as one from government agencies or CSIRO. More seriously, it was based on listing some of the individual small logging areas – the coupes – scheduled for cutting in 1995. The environment department added further coupes to give a total of 1297, which John Faulkner presented to David Bedall as needing protection from logging. The list made neither ecological nor industrial sense, because it is only the cumulative impact of many years' cutting across the landscape, not an individual coupe, that is significant, and it was politically disastrous because it put the Commonwealth in the position of licensing the operations coupe-by-coupe and year-by-year.

There was little love lost between those on the ALP's left wing, like John Faulkner, and those on its right wing, like the Resources Minister, David Bedall, who seems to have seen the list as an ambit claim which he could virtually ignore. Without taking the issue to Cabinet, he renewed the licences, increased one in Western Australia, granted a new one in Queensland, and protected only 85 coupes after objections. The environmental movement was outraged and the matter quickly escalated when his disregard of environmental advice was found by the Federal Court to have invalidated an earlier licence and cast doubt on the current ones.[27]

Faced with a major contest, the Prime Minister declared that 509 coupes would be reserved from woodchipping pending assessment, and that the licences would be reissued safe from legal challenge.[28] But this pleased neither environmentalists nor workers and the issue was drummed up as a crisis; 10 000 people protested at an environmental rally in Melbourne; reportedly as many as 6500 forest industry workers with their families and supporters blockaded Parliament House in Canberra for a week; five MPs threatened to resign from the Labor Party, and one Senator, John Devereux, did.[29] The Liberal State Premiers must have been delighted to see the Commonwealth Labor government in such disarray, and their forest services did nothing to defuse the issue by rescheduling logging to other areas while the assessments were carried out; it all had to be rushed under threats of mill closures and job losses. Federal Cabinet was left in the ludicrous situation of deciding whether particular batches of coupes, as they were hurriedly assessed, warranted protection or could be released for logging, and the livelihoods of working people were made needlessly uncertain.[30]

The long-term issues of making a transition to a sustainable future, both environmentally and socially, were addressed most seriously not by the federal Labor Party but by the NSW Branch as it contested the 1995 State election. It had spent two years consulting environmental groups and the major timber union to come up with a plan to replace logging of biologically rich old-growth forests with resources from plantations and regrowth, insist on firms adding more value in their processing as a condition of their forest licences, and provide substantial funds for retraining and restructuring.[31] Whether this proposal can be put into effect has yet to be seen, but it does attempt to take a higher road towards sustainable development.

High road

By some criteria, Australia has made considerable progress along what can be called the 'high road' to sustainable development. Four strengths stand out. First is the high level of scientific attention paid to the biodiversity of the comparatively small area of forests. Although the more that is discovered about their particular ecology and endemic species, the more formidable the future research tasks appear, Australia is at the level of international excellence in the work. Second, Australia also has strong research and technological skills for growing plantations and manufacturing wood products. Third, although a small nation, Australia has participated fully in the development of international environmental agreements. Last, perhaps surprisingly taken as an advantage in view of the turbulence of the last two decades, is the level of public interest, the intense involvement of the environmental movement, and some of the evolving processes to recognise and deal with the contests. Their very turbulence provides the impetus to consider sustainable development seriously.

With such strengths, the forest sector is in many ways well placed to follow the high road. Three-quarters of the forests still belong to the state, which makes their fashioning more amenable to public debate and policy than if they were in private hands. Public debate is lively. The right to participate in decision-making is not only keenly sought, but is gradually, though often grudgingly, being accepted. There are well-established agencies with cadres of professional staff which can manage national parks, state forests and plantations for whatever purpose is required. There is a fully-developed forest industry, much of it still mainly in Australian hands, with an experienced, if shrinking, workforce. If the high road is chosen, it can be followed competently.

It is also timely to do so. The long pioneering era of opening up the forests for the first time is almost at an end, as most forests have their basic roads and other infrastructure in place. They have been over-exploited in the past, which has forced a confrontation with the problems of sustained yield.[32] Few pristine areas remain outside national parks, and while their future will be fiercely contested, it will become clear one way or another by the end of the century. The refashioning of the forests accessible to the pulp and woodchip mills is perhaps half done, and the rate of cutting hardwood sawlogs has sharply declined. The future will depend on what can be called the 'second resource' of regrowth in the native forests and on the plantations.

Although it will always be relatively small in the Pacific region, the plantation estate of both pines and eucalypts can be expanded. Its productivity will increase as more areas that have been planted with improved strains and managed intensively come into production. The combined effects should enable domestic production to be increased so that, it might become economically efficient to reduce imports.

Submission and ignorance, or resolve and adaptability?

The choice of roads will never be easy, uncontested or entirely within the power of Australians to make, but it is not impossible, irreconcilable or totally expropriated either. It seems in the early 1990s as though Australia is poised, not knowing which road to take.

The low road has powerful interests and plausible advocates for leaving the market to choose. Submissiveness to market forces, freed from distortions and corrected for externalities, would be rewarded and intervention punished, or so the economic rationalists intone. The state should ensure this, they say, and streamline the bureaucracy to suit. But the real world is not so simple: there are more values than even perfect markets could recognise; powerful economic and political forces pursue their own ends; and Australia is a weak player on the periphery of a large region. If Australia is ignorant of the complexities and changes at work, it is likely to be their victim; it needs to be alert.

The high road is far more difficult to take, depending as it does on controlling short-term and private interests for long-term and public gain. No single criterion can mark a route through the several values and categories. Society and forest ecosystems are too complex for that; the route has to be scientifically understood and socially negotiated.

This has an inherent value: in recognising all the contesting viewpoints and participating in the management of the forests, the sense of community blooms.

The choice demands more than conjuring national resolve and political will to follow the high road. Rather than minimising the state, taking the high road would place the state, the locus of democracy and the guardian of the common wealth, as central. The state must be alert to the uncertain and threatening nature of Australia's future. It would need a lively, professional and well-staffed national bureaucracy to monitor resource changes, and to flag economic opportunities in the Pacific region. It would need equally lively State and regional agencies, to monitor resource and environmental changes, propose plans for transition towards sustainable development, encourage their debate, revise them frequently, manage the public forests, and facilitate the development of plantations.

There is much to be done now, and creating the future for Australian forests will become progressively more difficult. The resolve to undertake the tasks can be heartened by recalling von Mueller's injunction, written more than a century ago, with which this book began:

> Let us regard the forests as an inheritance, given to us by nature, not to be despoiled or devastated, but to be wisely used, reverently honoured and carefully maintained. Let us regard the forests as a gift, entrusted to any of us only for transient care, to be surrendered to posterity as an unimpaired property, increased in riches and augmented in blessings, to pass as a sacred patrimony from generation to generation.[33]

APPENDIX

LIST OF SPECIES AND TIMBERS

Common or trade name	*Botanical name*
Alpine ash	*Eucalyptus delegatensis*
Blackbutt	*E. pilularis*
Blue gum (Tas.)	*E. globulus*
Blush tulip oak	*Argyrodendron actinophyllum*
Brown barrel	*E. fastigata*
Brush box	*Lophostemon confertus*
Carabeen	*E. tesselaris*
Cedar, red	*Toona australis*
Celery-top pine	*Phyllocladus asplenifolius*
Cypress pine	*Callitris* spp.
Darwin stringybark	*E. tetradonta*
Flooded gum	*E. grandis*
Giant stinging tree	*Dendrocnide excelsa*
Grey gum (Central coast)	*E. propinqua*
Hinoki (Japan)	*Chamamecyparis obtusa*
Hoop pine	*Araucaria cunninghamii*
Huon pine	*Dacrydium franklinii*
Jarrah	*E. marginata*
Karri	*E. diversicola*
King Billy pine	*Atherotaxus selanginoides*
Lancewood	*Acacia shirleyi*
Leatherwood	*Eucryphia lucida*
Lilly-pilly	*Acmena smithii*
Mallet	*Eucalyptus occidentalis var. astringens*
Manna gum	*E. viminalis*
Marri	*E. calophylla*
Messmate	*E. obliqua*
Mountain ash (Vic.) or Swamp gum (Tas.)	*E. regnans*
Mountain grey gum	*E. cypellocarpa*
Myrtle beech	*Nothofagus cunninghamiii*
Paperbark	*Melaleuca* spp.
Pink bloodwood	*E. intermedia*
Red bloodwood	*E. gummifera*
Red gum, forest	*E. teretecornis*

Common or trade name	*Botanical name*
Red gum, river	*E. camaldulensis*
Red ironbark, narrow-leaved	*E. crebra*
Red stringybark	*E. macroryncha*
Sandalwood	*Santalum album, S. spicatum, S. lanceolatum.*
Sassafras	*Doryphora sassafras, Atherosperma moschatum*
She-oak	*Casuarina* spp.
Shining gum	*E. nitens*
Silvertop ash (Vic.) or Coast Ash (NSW)	*E. sieberi*
Smooth-barked apple	*Angophora costata*
Sugi (Japan)	*Cryptomeria japonica*
Tallowwood	*E. microcorys*
Tuart	*E. gomphocephala*
Wandoo	*E. wandoo*
Wattle	*Acacia* spp.
White box	*E. albens*
White mahogany, broad-leaved	*E. umbra*
White stringybark	*E. globoidea*
Yellow box	*E. melliodora*
Yellow carabeen	*Sloanea woolsii*

Source: D. J. Boland, et al. (1985).

NOTES

Introduction

1 This type of interpretation was developed by Massey (1984) for old industrial cities.
2 For interpretations which give the forests an active 'voice' in history, see Rolls (1981), Pyne (1991), and Griffiths (1992).
3 Barlow (1981); Hope and Kirkpatrick (1988); White (1986).
4 Macpherson (1983), pp. 1–4.
5 Bromley (1991), p. 21.
6 Young (1981, 1982).
7 See for example: List (1993); Eckersley (1992); Naess (1973); Devall and Sessions (1985).

1 Forests for home and empire

1 Hope and Kirkpatrick (1989).
2 Donaldson and Donaldson (1985). It is thought that the population increased rapidly some 4000–5000 years ago and that this coincided with the development of more complex societies which managed food resources more deliberately and held more frequent and elaborate ceremonies. Much of the material in this section is drawn from Boutland (1989).
3 Jones (1969); Lewis (1982).
4 Ellis and Thomas (1989).
5 Hanssen and Wigston (1989).
6 Palmer (1991).
7 Letter from the Committee of Seventeen (that directed the company from Holland) to the Governor-General (Van Diemen) and Councillors in Batavia, 11 April 1642, cited by Sharp (1968).
8 MacKnight (1976).
9 Hobsbawm (1952).
10 The mix of reasons that led to Australia being selected are canvassed in Martin (1978) and Frost (1994).
11 Roberts (1968).
12 Cox and Freeland (1969).
13 The task for sawyers was 1.4 cubic metres of sawn timber a week for experienced men and 1.1 for new convicts. Labourers could split shingles under a similar arrangement: Dargavel (1988b).

14 West (1981), p. 395.
15 Backhouse (1967).
16 Memorandum to Colonial Secretary, 14 October 1834 (State Archives of Tasmania CSO 1/732/15836/3). The form of application required particulars of the applicant's name, whether free or on ticket-of-leave, age, height, hair, eyes and complexion. In Victoria, recommendations from two respectable people were required.
17 Harris (1953) [1847] gives a colourful account of cedar cutting.
18 Two businesses were listed in the Sydney directories as 'Ship owner and timber merchant' in 1835, but by 1842 there were 21 listed as 'timber merchants'.
19 Linge (1979), p. 76.
20 Dargavel (1988a). See also the reference to Mathew's oxen-driven mill operating near Pymble from 1827 given by Hudson and Henningham (1986), pp. 223.
21 Although some 9000 convicts were later sent to Swan River and some to Port Phillip, most of them were already eligible for tickets-of-leave or conditional pardons.
22 *Maryborough Chronicle* (1988).
23 Jackson (1977), pp. 28–38.
24 Some timber transhipped in British ports is recorded as having come from the United Kingdom. Most of the North American wood was douglas fir from the west coast of the United States. It is still called 'oregon' in Australia, even though much of it comes from Canada. The kauri forests of the North Island of New Zealand contained huge trees which provided a light, very easily worked timber in particularly wide, clear boards. New Zealand also provided the odourless white pine, cut from kahikatea trees, that was used for butter boxes.
25 Patterson (1968), p. 45.
26 Industrialisation in Queensland was inhibited by the tariffs imposed on timber by New South Wales, though the rates were much less, and applied later than Victoria's: *ibid.*, p. 69.
27 Some of the city mills were large affairs with advanced woodworking machinery. See, for example, NSW (1870) and M'Leod (1868), p. 17.
28 Considerable research by members of the Light Railways Research Society is published in their journal, *Light Railways*, and monographs.
29 Houghton (1988).
30 The rights are detailed in a series of working papers: Dargavel, Goddard, and Caton (1986–87).
31 Carron (1985); Powell (1976).
32 The formal authority was usually given to the Governor-in-Council, which effectively meant the government of the day.
33 Carron (1985).
34 New South Wales, Department of Mines, Forests Branch, 1883, *Report.*
35 By providing the continuous motion of a circular saw with the constant cutting angle of a frame saw.
36 Based on British standard gauge: O'Dell (1956), pp. 15–16.
37 The teak forests of Burma (now Myanmar) and India (India, Pakistan and Bangladesh) and Thailand provide perhaps the prime example, but some Australian forests were drawn in too.
38 Mills (1988a).

39 For example, exports of jarrah, karri and tuart from the southern Sussex District in 1883–84 were sent to: South Australia 63 per cent, New Zealand 23 per cent, China 12 per cent, and Mauritius 2 per cent: WA, Legislative Council, *Votes and Proceedings*, 1884, paper A8. Import statistics from the other colonies are reported in Dargavel (1990).
40 Robertson (1959); WA, Legislative Council, *Votes and Proceedings*, 1870–71, paper A1.
41 Fall (n.d.).
42 Hamlin (1969).
43 Purcell (1988).
44 WA, Royal Commission on Forestry, Reports, in WA, *Parliamentary Papers*, 1903–4, paper no. 24; 1904, paper 11.
45 Dolin (1967).
46 Mills (1986).
47 Roche (1993).
48 Row (1980).
49 Calder (1980).
50 Mills (1989).
51 Beckett (1983).
52 *The Axemen's Journal and Sporting News*, July 1901.
53 *Ibid.*

2 Forests for industrialisation

1 Australia absorbed up to 43 per cent (1925–28) of Britain's entire foreign lending to public authorities: Cochrane (1980).
2 Mussen (1919).
3 Watson and Cohen (1969); Benjamin (1959); Jeffreys (1947); Rawson (1953).
4 It was built on the site of the Huon Timber Company's sawmill at Kermandie.
5 Mayer (1964).
6 The only reference to this contract that I have been able to find was given by W. Bunston (joint manager of the war-time newsprint pool) in an untitled article in ANM's house journal, *Newsprint Log*, 1(1), 1945.
7 Kraft pulping is an alkaline chemical process that produces paper with strong burst and tear characteristics. The mills are usually large, and incorporate recovery systems for the chemicals.
8 Sinclair (1990).
9 Benjamin (1966); Johns (1962).
10 *Jobson's year book of public companies*, 1964.
11 *Mercury*, 24 September 1964.
12 On the site of the old Kermandie pilot pulp mill and the Huon Timber Company's old sawmill.
13 The proposers were Holyman's, a Tasmanian shipping company with interests in sawmilling, K. D. Atkins, a Launceston timber merchant and sawmiller, and the network of mainland timber merchants – Gunnersens, Le Messurier and Alstergren – represented in Kilndried Hardwoods.

14 Australia, Tariff Board (1954), *Report on paper and paper products*, cited by Davidson and Stewardson (1974).
15 Russ and Tanner (1978), p. 90.
16 APPM, *Proceedings Annual General Meeting*, 1946.
17 APPM built a second hardboard mill at Ipswich in Queensland in 1958. It combined its hardboard interests with those of Colonial Sugar Refiners Ltd (CSR) to form Hardboards Australia in 1967, which eventually closed the Burnie mill in 1978.
18 Davidson and Stewardson (1974); Johns (1962).
19 Sulphite pulping is an acid-based chemical pulping process. Normally, its chemicals are not recovered and its effluents are polluting.
20 Tasmania, *Wood-Pulp and Paper Industry Encouragement Act 1924*; *Wood-Pulp and Paper Industry Encouragement Act 1926*; *Associated Pulp and Paper Mills Act 1936*.
21 The rate was increased from 1 shilling per cord ($0.03 per cubic metre) by 5 per cent for every 1 per cent that the company exceeded 8 per cent, up to the maximum of 2 shillings, 6 pence ($0.07 per cubic metre).
22 Tasmania, Conservator of Forests to Minister Controlling Forestry, 25 November 1924, Tasmanian State Archives FC 5/5459.
23 Victoria, *Wood Pulp Agreement Act 1936*.
24 Three-quarters of the pulpwood had to be from *Eucalyptus regnans* or *E. gigantea* and of this one-third had to be *E. regnans* under 60 years of age.
25 Dixson (1963).
26 Lewis (1988).
27 Of the volume of logs going into a typical hardwood sawmill, about 45 per cent ends up as saleable timber, 10–12 per cent as sawdust, 30–35 per cent as usable offcuts suitable for chipping. The remainder and the sawdust is burnt, sometimes generating heat for kilns.
28 Algar (1988).
29 M. R. Garrett, 'Comments on an intensive checking system of sawmilling operations in the North-West Division', 1937, State Archives of Tasmania FC4/ASS5/5.
30 Sawmillers who had been cutting in the area dedicated as the Wollemi National Park in 1980 were unsuccessful in claiming compensation for the loss of expectation of renewal.
31 Mann (1958).
32 Lawrence (1963).

3 Forests for conservation and delight

1 Victoria, Report on the advisableness of establishing state forests, *Votes and Proceedings of Parliament*, 1864–65.
2 Victoria, Papers relating to forest conservancy, *ibid.*, 1874.
3 Queensland, Report on the Royal Botanic Garden, *Votes and Proceedings of the Legislative Assembly*, 1875.
4 Colonial attitudes were reflected in the way rainforests were referred to as 'brush' in New South Wales, and as 'scrub' or 'jungle' in Queensland.

5 See Powell (1976), pp. 46–8, 68–74. Mueller's work was recognised by a knighthood in Britain and a barony in his native Germany.
6 Lucas (1988). Royal Societies were formed in Van Diemen's Land, Victoria and New South Wales in 1844, 1860 and 1866 respectively from societies formed a few years previously. South Australia and Queensland followed in 1880 and 1884. A Royal Society was not formed in Western Australia until 1913.
7 For example, Ronald Gunn, who held various official positions, such as Police Magistrate in Launceston, became a most knowledgeable botanist who contributed extensively for the herbarium at Kew.
8 Lewis (1975).
9 Laslett (1875).
10 Kohlsted (1983).
11 Sydney International Exhibition (1881), p. lxxxvii.
12 Warren (1911); Julius (1906).
13 Bolton (1981), ch. 9.
14 South Australia, *National Park Act 1891*, cited by Black and Breckwoldt (1977).
15 Maiden and Campbell (1895–1908).
16 J. E. Brown (1896), Report on the forests of Western Australia, their description, utilisation and proposed future management, WA, Legislative Council, *Votes and Proceedings*, 1896, paper 27. He was the son of a Scottish forester, James Brown, and was the most experienced forester in the colonies at the time, having previously been Conservator in South Australia and Director-General in New South Wales.
17 Maiden (1897).
18 Maiden inserted these comments with parts of his 1897 Anniversary Address in the middle of a description of a little-known tree, *Endriandra muelleri*, in his flora of NSW: Maiden and Campbell (1895–1908), part xxxi (1908), pp. 113–15.
19 O'Shaughnessy (1986).
20 Ellery to Commissioner of Public Works, 13 April 1880, cited by O'Shaughnessy (1986).
21 Queensland, Director of Forests, *Report 1918*, cited by Carron (1985), p. 101.
22 Schlich (1982) [1922].
23 The unpublished report is cited by Moulds (1991), p. 18.
24 G. S. Perrin was Tasmania's first Conservator of Forests in 1886–88.
25 B. Ribbentrop, Report on the State Forests of Victoria, in Victoria, *Papers Presented to Parliament 1896*, vol. 4, paper no. 42.
26 Victoria, Royal Commission on State Forests and Timber Reserves 1898–1901, *Progress Reports 1–13*; and final report: *Forestry in Victoria, Australia: legislation, control, and management requisite; with some account of the other colonies of Australasia and of forestry in other countries*, Govt Printer, Melbourne.
27 Hutchins (1916). His scathing, if idiosyncratic, report extended to all States.
28 Carron (1985).
29 Black and Breckwoldt (1977).
30 Carron (1985), pp. 120–3.

31 H. R. Mackay, Victoria's Conservator of Forests, referring to the 1911 Conference. *Report of the proceedings of the* [2nd] *Interstate Conference on Forestry, Melbourne, November 1912.* Govt Printer, Melbourne.
32 Report of the proceedings of the [1st] *Interstate Conference on Forestry,* Sydney 1911. Govt Printer, Sydney. 1912.
33 Report of the inaugural meeting of the Empire Forestry Association, 1920, *Empire Forestry,* 1: 3–10.
34 Obituary, 1971, *Commonwealth Forestry Review,* 70: 1–3.
35 *Proceedings (5th) Interstate Conference on Forestry,* 1920.
36 Steane (1937).
37 Hutchins (1916), pp. 193–6.
38 Forests Commission of Victoria, *Annual Report 1930–31.*
39 Pyne (1991).
40 Moulds (1991), p. 6; Griffiths (1992).
41 Stretton (1939).
42 McGrath (1957).
43 Hutchins (1916), p. 108.
44 NSW, *Report of the Department of Forestry 1916,* Govt Printer, Sydney.
45 Lennon (1993).
46 Jolly (1920).
47 Jacobs (1955).
48 Barrère, Heisch and Lerat (1962), ch. 6; Reed (1954).
49 Carron (1990).
50 Lewis (1988).
51 See Carron (1985), pp. 244–69, for events leading to the Australian Forestry School.
52 A special exemption was made for Victorian foresters graduated from Creswick for a few years.
53 Dargavel (1980); Hawkins (1987).
54 Bolton (1981), ch. 12.
55 Stretton (1939), p. 5.
56 Robin (1991).
57 Pepper (1984).
58 Objects of the Mountain Trails Club, as revised in 1924, are cited by Johnson (1974), p. 18.
59 Carron (1980).
60 The loans to the States for 35 years were free of interest and repayments for the first 10 years.
61 Carron (1985), pp. 307–12.
62 Mann (1990).
63 APM also started plantations in Queensland and New South Wales, but sold them when proposed pulp mills did not eventuate.
64 Hindmarsh (1983?); Wells (1978?).

4 Forests for Japan

1 Sinclair (1987).
2 Garnaut (1989), Table 2.4.

3 The proportion of manufactures in total imports increased from roughly one-third to one-half between 1985 and 1989: *ibid.*, Table 3.7.
4 Osako (1983); Izumi (1988); Totman (1989, 1993).
5 Kumazaki (1988).
6 Ishii (1991).
7 Apart from a period in the 1920s when North American timber was brought in, under reduced tariffs, for rebuilding after an earthquake: Kumazaki (1988).
8 Sugahara and Ouchi (1988); Ouchi (1988).
9 The production of firewood and charcoal at the end of the 1970s was only 3 per cent of its level in the 1950s.
10 Mitsuda (1984).
11 Ishii (1991); Penna (1992).
12 Young (1979); Kojima and Ozawa (1984).
13 Penna (1992).
14 *Pulp and Paper Perspective*, April 12, 1992.
15 Including allowance of approximately 40 per cent for the higher yield of pulp per cubic metre from hardwoods than from softwoods.
16 Report of statement by A. G. Hanson at ACF Symposium on the Conservation of the Australian Coast, *ACF Newsletter*, January 1970, p. 5.
17 Australia, Working Group on the Economic and Environmental Aspects of the Export Woodchip Industry (1975).
18 Benjamin (1959).
19 The royalty rates were set on complex formulas designed to put all sawmillers on an equal footing in the market irrespective of where they operated. The formula gave a negative sum for the Eden area, but a token minimum rate was charged.
20 The Commission's FORINS survey was conducted in 1971–72.
21 Dillinghams, Brambles and Australian Paper Manufacturers, Harris–Daishowa (Aust.) Pty Ltd, *Annual Report*, 1975.
22 Friends of the Earth, Japan (n.d.), Report on Japanese import of Australian woodchips and Daishowa Paper Manufacturing Co., mimeo.
23 Daishowa also bought a paper mill in the western United States and a forest products company in Quebec, and started to produce for markets outside Japan.
24 New South Wales, Pulp and Paper Industry Task Force (1989).
25 McCuaig (1981).
26 The original members of TPFH were: Risby Bros Pty Ltd, W. A. McKay Pty Ltd, Kilndried Timber Industries Ltd, B. G. Clennett Pty Ltd, Kemp and Denning Ltd, Tasmanian Board Mills Ltd, Chesterman and Co Pty Ltd, IXL Timber Pty Ltd, Hall Bros, S. T and Y. C. Brown, R. J. Hanlon, Fehlberg Timbers Pty Ltd, and Crisp and Gunn Co-op Ltd.
27 TPFH, *Annual report*, 1975–78; *Australian Timber Journal*, 1968, 34(11): 70–1; 1971, 37(6): 15; 1973, 39(11): 57.
28 *Jobson's year book of public companies*, 1964–72; *Australian Financial Review*, 26 September, 8 October 1968.
29 *Mercury* 5, 12, 21 June 1968; *Examiner*, 22, 29 June 1968.
30 Chapter 9.
31 Including the 25 per cent holding of British and American Tobacco.
32 Chapter 8.

33 *Australian Financial Review*, 5, 8 January 1970.
34 Interview with Mr G. Harrison, General Manager, Northern Woodchips, 1 November 1979.
35 NSW, State Pollution Control Authority (1975).
36 *Age*, 3 June 1981; *Geelong Advertiser*, 7 October 1981.
37 Streeting and Imber (1991).
38 Byron and Douglas (1981); Parker (1986); Streeting and Imber (1991).
39 Australia, Working Group on the Economic and Environmental Aspects of the Export Woodchip Industry (1975); Tasmania, Forestry Commission (1978a).
40 Chapter 9.
41 Tasmania, Forestry Commission (1978b).
42 Australia, Senate Standing Committee on Science and the Environment (1977), p. 199.

5 Forests for work and regional development

1 Hutchins (1916), pp. 200–7; Australia, Development and Migration Commission (1929).
2 Windschuttle (1979); Langmore and Quiggin (1994).
3 Pusey (1992).
4 Powell (1988), pp. 262–4.
5 Simpfendorfer (1962).
6 RAC (1991).
7 The 'Resawn and dressed timber' industry category (ASIC 2532) is included, even though part of this industry depends on imported timber. 'Pulp, paper and paperboard' (ASIC 2631) depends partly on imported pulp. 'Logging' (ASIC 0303) does not include log and finished product transport, which is considered to be in the 'Transport' industry. CSIRO logging surveys recorded 1052 drivers in 1981 and 887 in 1992: McCormack and Sebire (1983); McCormack, Roberts, and MacArthur (1992).
8 The data problems are discussed in Dargavel, Johnston, and Boutland (1985).
9 FAFPIC (1987).
10 In this section, I am indebted to Jacquie Tracey for advice drawn from her research in progress: Coping with insecurity: family logging firms in northern New South Wales, PhD thesis, ANU. See also Dargavel, Parker and Tracey (1992).
11 Where large infrastructure costs have been required, companies have sometimes run their logging operations directly, as in the cases of the first large sawmills in Western Australia and ANM in the Florentive Valley.
12 Marx's exposition of piece-wages can be extended to the contractors: Marx (1954) [1st English edn 1887], vol. 1, ch. 21.
13 Statistical reports generally combine logging with forestry. Together, they have a slightly better safety record than mining, but are markedly worse than that in the wood and paper manufacturing: Australia, National Occupational Health and Safety Commission (1993).
14 Originally the Imlay Timber Contractors Association.

15 Parker, Dargavel and Ord (1992).
16 Some foresters would be recorded by ABS under various industry, research and other categories rather than 'forestry'. Statistics in this section are drawn from Neville and Ellis (1989).
17 See Chapter 10.
18 McKinnell (1993).
19 Parker, Dargavel and Ord (1992).
20 McCormack, Roberts and MacArthur (1992).
21 Victoria, Task Force (1983).
22 Forestry Commission of NSW, *1991–1992 Annual Report.*
23 *Australian Forest Resources,* 1990 and 1991, Data for 1990–91.
24 Data for 1988–89. Production from *Australian Forest Resources*; employment from ABS, *Manufacturing Establishments,* cat. nos. 8203, 8209.4.
25 Ivor Ries, *Australian Financial Review,* 10 June 1992.
26 *Appita,* 1994, 47(2): 90; *Australian Financial Review,* 26 January 1994.
27 *Appita,* 1994, 47(2): 90.
28 1.7 rather than 7 people per 1000 tonnes produced.
29 *Sydney Morning Herald,* 27 June 1994; *Appita,* 1994, 47(3): 192.
30 *Appita,* 1989, 42(2): 89; 42(5): 325.
31 Dargavel (1990).
32 Ord and Dargavel (1989).
33 For the forest sector, this view was advanced and later repudiated by the famous forest economist, Jack Westoby (1987).
34 Reilley (1974).
35 Kennedy and Lodge (1976).
36 A very substantial five-year study into the effects of forest developments in Oberon, NSW, was commissioned by the Forestry Commission of New South Wales: Dwyer Leslie (1990–93).
37 See particularly Dempsey (1990, 1992) and Wild (1974).
38 Watson (1990), pp. 51–7.
39 Elton (1976).
40 Watson (1990).

6 Forests for existence and experience

1 From the extensive literature, see for example: Naess (1973); Bookchin (1982); Duval and Sessions (1985); Pepper (1984); McCormick (1989); Eckersley (1992); Knudtson and Suzuki (1992); Merchant (1992); List (1993); Porritt and Winner (1988).
2 Offe (1985).
3 Sinden (1981); Mercer (1981).
4 ABS (1986).
5 Turner (1981).
6 Clay, Hingston, and Aslin, (1988).
7 Turner (1981).
8 Public employment rose from one-quarter of the workforce in the 1950s to nearly one-third by the mid-1980s, reduced to 29 per cent by 1991: Reserve

Bank of Australia, *Australian Economic Statistics*; ABS, *Employed wage and salary earners*, cat. no. 6248.0.

9 Commonwealth and State governments held 37 inquiries into forestry and forest products between 1971 and 1990, and a far greater number of regional land-use and forest management planning studies.

10 Dumaresq (1988).

11 Powell (1988), pp. 234–45.

12 Dargavel and Ferguson (1975); Taylor (1975).

13 Gee and Fenton (1978); Jones (1981); Green (1981); Alexander (1981); Thompson (1984); Easthope and Holloway (1989); *ACF Newsletter*, 1973, 5(5); *Bulletin*, 3 November 1973.

14 *ACF Newsletter*, 1972, 4(4).

15 Alexander (1981).

16 Easthope and Holloway (1989).

17 Thompson (1984).

18 Richardson (1990); Australia, Committee of Inquiry into the National Estate (1975).

19 Rosauer (1993).

20 Australia, *Environment Protection (Impact of Proposals) Act 1974–75*.

21 NSW, *Environment Planning and Assessment Act 1979*; Preston (1989).

22 Bonyhady (1993), pp. 71–8.

23 Australia, Committee of Inquiry into the National Estate (1975); ACF (1982).

24 Doyle (1990).

25 The centre in which this book was written has been associated with the ACF most notably though the centre's first director, Professor Frank Fenner, who was vice-president of the ACF in 1971–73, and through Dr H. C. (Nugget) Coombs who was president of the ACF in 1977–78.

26 *ACF Newsletter*, 1967, 1.

27 R. D. Piesse (Director, ACF), Address, in *ACF Newsletter*, 1970, 2(3):16; H. R. H. Prince Phillip, The Duke of Edinburgh, Conservation's real message, *ibid.*, 1972, 4(3): 2.

28 *ACF Newsletter*, 1973, 5(5); *Bulletin*, 3 November 1973.

29 Trish Caswell took over as director in 1992.

30 *Business Review Weekly*, 17 November 1989.

31 Holloway (1986); Green (1981).

32 Jarrott (1980).

33 Johnson (1974).

34 Johnson (1982).

35 Australian Council of National Trusts (1993); ABS, *Year Book Australia, 1991*, pp. 318–22.

36 Australia, Committee of Inquiry into the National Estate (1975).

37 J. Tedder, North Coast Environment Council, pers. comm., 18 June 1993.

38 See Chapter 8.

39 See Chapter 7.

40 Major (1991).

41 See Chapters 7 and 8.

42 Mueller (1984).

43 Specht, Roe and Broughton (1974).

44 Frankel (1975).

45 Australia, Senate Standing Committee on Science and the Environment. (1977), p. 12.
46 RAC (1993).
47 Lothian (1994); Bean (1994); Papadakis (1993, 1994).
48 Papadakis (1993), Table 5.8, based on 1990 Australian Election Study.
49 McAllister (1991). The survey collected responses from 2179 people nationally, including 678 people in the Eden–Monaro and Gippsland electorates.
50 SSDA (1992).
51 ABS (1993), Table 5.1. showed that the media, newspapers or TV were the most common source of environmental information which reached 86 per cent of the population.
52 I am indebted to Roger Green for his insights on the media.
53 *Weekend Australian*, 31 March 1–April 1990.
54 Thorough media studies are needed to establish or refute this assessment.
55 *Age*, 13 May 1986.
56 *Age*, 30 May 1986; *Sydney Morning Herald*, 20 April 1987, for examples.
57 Lunney and Moon (1987).
58 Jones et al. (1993), Table F1.
59 ANOP Research Services (1994).

7 Contesting existence, ecology and experience

1 Working Group on the Economic and Environmental Aspects of the Export Woodchip Industry (1975). vol. 1, p. 24.
2 Routley and Routley (1975).
3 *Ibid.*, pp. 1, 219.
4 ACF (1975); Jones (1975); Campaign to Save Native Forests (WA) (1975).
5 Australia, Senate Standing Committee on Science and the Environment (1977), p. 4.
6 Australia, Working Group on the Economic and Environmental Aspects of the Export Woodchip Industry (1975).
7 Heyligers (1975, 1977).
8 Carter (1975).
9 Australia, Senate Standing Committee on Science and the Environment (1977) pp. 6, 16.
10 Tasmania, Forestry Commission, *Annual Reports*, 1971–72 to 1979–80.
11 Board of Inquiry, Private forestry development in Tasmania, *Tasmania, Journals and Printed Papers of Parliament*, 1977, paper no. 25.
12 State EISs were also prepared for woodchip proposals in Western Australia (1973) and New South Wales (1975).
13 W. D. Scott and Co. (1977).
14 *Ibid.*, p. 42.
15 Harris–Daishowa (Australia) Pty Ltd (1977).
16 *Ibid.*, p. 32.
17 I am indebted to John Formby for his contribution to this section.
18 Tasmanian Woodchip Export Study Group. (1985a, 1985b).
19 Centre for Regional Economic Analysis (1985). Its errors were corrected in the Supplement to the EIS.

20 Harris–Daishowa (Australia) Pty Ltd (1986a).
21 New South Wales, Forestry Commission (1986).
22 Harris–Daishowa (Australia) Pty Ltd. (1986b).
23 The police perspective is given by Hinchey (1991).
24 *Age*, 13 June 1989.
25 *Australian Financial Review*, 4 January 1989; *Sydney Morning Herald*, 22 February 1989.
26 *Sydney Morning Herald*, 11 April 1989.
27 Tasmania, *Forest Practices Act 1985*.
28 Norton and May (1993).
29 See review by Lindenmayer (1993).
30 A paper on 'The pine planting program in NSW' was widely circulated prior to the publication of *The fight for the forests*.
31 For recent assessment, see Flinn and Turner (1990).
32 Australia, House of Representatives Select Committee on Wildlife Conservation (1972).
33 For summary of position at end of the 1970s, see Native Forests Action Council of Monash University (1981).
34 ACF (1973), *Softwood forestry and conservation*, Policy statement no. 17, mimeo.
35 Pratten (1983).
36 Flinn and Turner (1990); Brown (1990); Shepherd et al. (1990); Cromer et al. (1977).
37 RAC (1991), vol. 2, p. G3, cited 34 categories of terrestial reserves. Data for 1988 from Mobbs (1989).
38 Includes various forms of conservation reserve and parks from which wood production is excluded: *Australian Forest Resources* (1976, 1991). The comparison may be affected by a change in the classification system and errors in estimating the amount of forest.
39 Hall (1988) examined the 'worthless lands' hypothesis.
40 Goldstein (1979). Western Australia and Victoria separated the parks and wildlife functions.
41 *Herald*, 4 May 1977.
42 NSW, Forestry Commission, *Annual Report* 1989–90.
43 The other World Heritage areas are: Willandra Lakes (1981), Great Barrier Reef (1981), Lord Howe Islands (1982), Kakadu (1981/1987), Uluru (1987) and Shark Bay (1991).
44 Mercer (1991), pp. 102–10; Carron (1993).
45 Gee and Fenton (1978).
46 Thus providing the constitutional lever for the Commonwealth to stop the Franklin dam.
47 Australia, Commission of Inquiry into the Lemonthyme and Southern Forests (c. 1988).
48 See Chapter 9.
49 Bonahady (1993), pp. 128–46.
50 Australian Heritage Commission, *Annual Report 1992–93*.
51 Some places fell in more than one category.
52 RAC (1992), vol. 2A, pp. 110–11.
53 Purdie and Cavanagh (1993).
54 See Chapter 10; Rossauer (1993); Australian Heritage Commission and Department of Conservation and Natural Resources, Victoria (1994).

55 Mosley (1978).
56 ACF (1975); Washington (1984).
57 Helman et al. (1976).
58 It chose a minimum size of 25 000 hectares with a similar buffer zone, a core area free from major indentations, and a minimum width of 10 kilometres.
59 Lesslie, Taylor and Maslen (1993), pp. 1–2.
60 LCC (1991).
61 The major scientific text is Adam (1991) who gives substantial treatment to the contentious question of definition. An account of the environmental protests is Kendell and Buivids (1987).
62 Frawley (1991). In North Queensland, only cedar was cut in 1885, 10 species in 1900 and at least 124 in 1986: Vanclay (1992).
63 Anderson and Shea (1991).
64 ACF (1977).
65 Werren and Kershaw (1987, 1991), first published by WWF in 1984.
66 Australia, Working Group on Rainforest Conservation (1985).
67 Colong Foundation (1983); *Habitat*, 1976, 4(3), Special issue on the Border Ranges.
68 Watson (1990).
69 Adam (1987).
70 Mercer (1991), pp. 110–23.
71 Frawley (1987).
72 Winter and Atherton (1987); Young and McDonald (1987).
73 Rainforest Conservation Society of Queensland (1986).
74 ERIN (1994).
75 Falconer (1984).
76 Kendell and Buivids (1987).
77 Victoria, Dept of Conservation, Forests and Lands (1988).
78 Tyler (1979). Legislation was passed in WA 1980 (flora only), Victoria 1988, New South Wales 1991, and Queensland 1992.
79 ERIN (1993).
80 Phillips (1990). Although the mainland populations of koalas were of concern in Victoria, those on the islands of Westernport Bay seemed secure and prevented the koala being classified in the national list. Koalas do not occur naturally in Tasmania or Western Australia.
81 Lindenmayer, Tanton and Norton (1990); Lindenmayer (1994).
82 Claridge (1993).
83 Neave et al. (1994).
84 Walker and Bridgewater (1992).

8 Contesting industries and regions

1 Gunnersen (1977).
2 ACF, Forestry in Australia, Policy Statement No. 36, 10 July 1983; Wooten (1987).
3 B. Gibson, President's Report, *NAFI Annual Report 1987–1988*, p. 3.
4 Interview, Mr Paul Edwards, Melbourne, 23 July 1987.

5 Flo Paynter of Bombala Timber Support Group, *Canberra Times*, 20 February 1994.
6 Johnson (1974).
7 Moulds (1991), ch. 10.
8 Oates (1982).
9 McKinty (1969); Moulds (1991), ch. 10.
10 Ferguson (1985).
11 Victoria (1986).
12 East Gippsland Coalition (1985).
13 *Age*, 11, 17 January, 2, 3, 4 February 1984.
14 *Age*, 3, 4 February 1984.
15 *Age*, 3 March 1984. It also called for a new Occupational Health and Safety Act.
16 Christoff and Blakers (1986).
17 Stanley and King (1986).
18 Age, 29 March 1986.
19 *Potoroo Review* (Newsletter of the Concerned Residents of East Gippsland), 1994, no. 138.
20 Smith and Woodgate (1985).
21 See Chapter 8; Lindenmayer (1994).
22 Because fast-growing young trees regenerated on the cut-over sites 'drink' more water than the slow-growing older ones they replace.
23 Forestry Commission of NSW, *Annual Report 1983–1984.*
24 Gibbs (1992).
25 Stein J., Land and Environment Court of NSW, 1991, Corkill v. Forestry Commission of New South Wales [No. 2] 1991, *Local Government Reports of Australia*, 73, p. 160.
26 Bonyhady (1993), pp. 95–103; NEFA (1993). Notes for Chaelundi State Forest field trip, 7 February 1993. Inaugural Maurice Wyndham Conference 'Sustainable Forestry in Australia: Future Directions', Armidale: University of New England, mimeo.
27 Bonyhady (1993), pp. 95–103; *Environment Planning and Law Journal*, 1992, 9(1): 1–3.
28 Analysis by Gloria Jackson, CRES, June–July 1994.
29 RAC (1992), vol. 1, pp. 137–53.
30 *Ibid.*, p. 152.
31 *Ibid.*, vol. 2b, p. T8; FICA [1988].
32 *Australian* and other major newspapers, 4 March 1991.
33 All levels of government ran programs during the mid-1980s and a National Advisory Group on Local Employment Initiatives was established. Many schemes are described in the magazine, *Work Matters*, published 1986–89.
34 Tarlo and Miller (1985).
35 Tasmanian Woodchip Export Study Group (1985).
36 Victoria (1986).
37 Australia, Industry Commission (1993), p. 16.
38 RAC (1991), vol. 2: W1–18.
39 Bonyhady (1993), p. 138.
40 FAFPIC (1984).
41 Jaakko Pyory (1986).

42 FAFPIC (1985).
43 FAFPIC (1987).
44 Cameron and Penna (1988), which followed a critique of the industry in ACF (1987).
45 FAFPIC (1989).
46 The proceedings of a national conference to review the information were published in Dargavel and Semple (1990).
47 Areas as at 1991.
48 Boutland, Byron, and Prinsley (1990); Hall (1990); Loane (1990).
49 Dargavel (1990).
50 Australia, National Plantations Advisory Committee (1991).
51 Stephens, Hansard, and Dean (1993).
52 *IFA Newsletter*, 1993, 34(3); *Appita, 1994*, 47(1).
53 Sinclair Knight and Partners (1988).
54 *Globe and Mail* (Canada), 5 May 1988.
55 Milne (1990); Chapman (1992).
56 *Australian*, 15 February 1989.
57 J. Button, Senator Press statement, 18 August 1988.
58 CSIRO, *Paper on the Wesley Vale pulp mill*, 13 March 1989, mimeo.
59 *Australian*, 18 March 1989.
60 *Age*, 16 May 1975.
61 Victoria (1986), p. 36.
62 Victoria, Dept of Conservation, Forests and Lands (1988).
63 Gruen, Leslie and Smith (1989).
64 Victoria, Dept of Conservation Forests and Lands, East Gippsland Paper Products Development Study (1989).
65 Senator J. Button, Senator G. Richardson, and J. Kerrin, *Pulp and paper industry package*, Ministerial statement; Nelson and Crossland (1992).
66 *PR Newswire*, 10 June, and *Financial Post*, 15 June 1994 (Reuter News Service).
67 *Australian Financial Review*, 4 February 1993.
68 *Ibid.*, 23 August 1993.
69 Hall (1988).

9 Contesting politics and the state

1 Kellow (1987).
2 See Chapter 6.
3 Haward and Smith (1990).
4 Papadakis (1993), Table 6.3.
5 *Ibid.*, pp. 187–9.
6 Greenland (1989).
7 See Chapter 7.
8 Tasmanian Trades and Labor Council [1989].
9 Interview, Bob Richardson, ACTU, Melbourne, 20 July 1987.
10 Davis (1986).
11 Haward and Smith (1990).
12 *Tasmanian Parliamentary Accord*, 29 May 1989; Pybus (1990).
13 *Conservation News*, 1990, 22(9): 7.

14 Tasmania, *Public Land (Administration and Forests) Act 1991.*
15 Warhurst (1990); Doyle (1990).
16 *Park Watch,* June 1988, 158: 18–20.
17 Bean (1994).
18 Papadakis (1994).
19 Environment was included in changing aggregations such as the Dept of Environment, Sport and Territories, 1993.
20 Davis (1987).
21 Australia, Forestry Branch, Dept of Primary Industries and Energy (1990), *South-east forests of NSW: analysis of wood resource and forest management considerations in the context of the Australian Heritage Commission Act 1975,* mimeo, 12 pp.
22 Fitzgerald (1987).
23 Batini (1987).
24 See Chapter 8.
25 Bonyhady (1993), pp. 83–103.
26 In New South Wales the Forestry Commission's *Indigenous forest policy* (1976) and *Wildlife policy* (1980) set up the system of Preferred Management Priority (PMP) classification, flora and fauna surveys, and making operational prescriptions which took specific account of the environmental and other factors on each part of the forest: Forestry Commission of NSW, *Annual Report,* 1984–85.
27 Bonyhady (1993).
28 NSW, *Timber Industry (Interim Protection) Act 1992 No. 1.*
29 NSW, Parliament, Public Accounts Committee, 1990, *Report on the Forestry Commission* (Report no. 52).
30 Evatt Research Centre (1989).
31 Pusey (1991).
32 Wilson (1979).
33 Benson and Mutton (1988).
34 RAC (1992), vol. 1, p. 454.
35 Evans (1990). New Zealand's native forests were transferred to the Department of Conservation, with only a few small logging operations being allowed to continue.
36 *Newsletter* (IFA) September 1993, 34(3): 18–19.
37 Carron (1993).
38 Norton, Mackey, and Lindenmayer (1990).
39 See Chapter 9.
40 *Resource Assessment Commission Act 1989* (Cwlth), Schedule 1.
41 '"Environment" includes all aspects of the surroundings of human beings, whether affecting human beings as individuals or in social groupings', *ibid.*, sec. 3.
42 RAC (1991, 1992).
43 Australia, Ecologically Sustainable Working Groups (1991), Appendix A, Charter letter from Prime Minister, 29 August 1990.
44 Stewart and McColl (1994).
45 *Sydney Morning Herald,* 14 July 1989.
46 Joint Scientific Committee (1990).
47 Laughlin (1992).
48 Victoria, Government (1986).
49 Brinkman (1990).

50 Victoria, Dept of Conservation and Environment (1990–92).
51 *Age*, 1 August 1992.
52 See papers in Gardner (1993).
53 *Australian*, 2 November 1990.
54 Fisher (1993).
55 Australia, *Intergovernmental agreement on the environment*, 1 May 1992.
56 Australia (1992).
57 *Ibid.*, 11–12.
58 Australia, Commonwealth Inter-departmental Committee on Regional Forest Agreements (1994).

10 Forest prospects in a changing world

1 A good world overview is given by Bazett (1993); for an Australian overview see Dargavel (1992).
2 Potter (1993).
3 Budowski (1988).
4 For example, Malaysian Timber Association decided to reduce log exports by 15 per cent: *Nikkei News Bulletin*, 13 March 1991.
5 District Court Judge W. Dwyer, *USA Today*, 24 May 1991.
6 Bazett (1993).
7 *Reuter News Service*, 14, 20 June 1994.
8 Since Amcor took over making printing and writing paper from APPM in 1993.
9 Amcor, *Annual Report*, 1994.
10 Roche (1992).
11 *Australian*, 5 November 1994.
12 Daio Paper Corporation and Itochu Corporation announced that they would form a joint venture with the Shanghai Paper Corporation, which also has a joint venture with Daishowa: *Reuters News Service*, 8 July 1994.
13 Specht, Roe and Boughton (1974).
14 Hamilton (1993), for example.
15 Köstler (1956), p. 397.
16 Society of American Foresters (1993).
17 Australia, Industry Commission (1993).
18 Langmore and Quiggin (1994).
19 Population Summit of the World's Scientific Academies (1993).
20 Boyden (1993).
21 Brown (1992).
22 The fall of communism and the 'fallacy of the single mind' is discussed by Blackburn (1991).
23 Fukuyama (1992).
24 United Nations, World Commission on Environment and Development (1987).
25 See for example Dovers and Handmer (1993).
26 Walker (1992).
27 *Sydney Morning Herald*, 23 January 1995.
28 *Canberra Times*, 28 January 1995.
29 *Canberra Times*, 30 January, *Australian*, 3 February 1995.

30 *Canberra Times*, 6, 7 March 1995.
31 *Sydney Morning Herald*, 9 March 1995.
32 Ludwig, Hilborn and Walters (1993).
33 Mueller (1879), p. 275.

BIBLIOGRAPHY

ABS (1986). *Environmental issues and usage of national parks, Australia.* Cat. no. 4115.0.

ABS (1993). *Environmental issues: people's views and practices.* Cat. no. 4602.

ACF (1973). *Softwood forestry and conservation.* Policy statement no. 17. Mimeo.

ACF (1975). *The great forest sellout.* ACF, East Melbourne.

ACF (1975). *Wilderness conservation.* ACF, East Melbourne.

ACF (1977). *The rainforests: ACF viewpoint.* ACF, East Melbourne.

ACF (1982). *The green pages: directory of non-government environmental groups in Australia.* ACF, Melbourne.

ACF (1987). *Australia's timber industry: promises and performance.* ACF.

Adam, P. (1987). *New South Wales rainforests: the nomination for the World Heritage List.* National Parks and Wildlife Service of NSW.

Adam, P. (1991). *Australian rainforests.* Clarendon, Oxford.

Alexander, K. (1981). The national stage. In Green (1981).

Algar, W. H. (1988). Forestry and forest products. In Australian Academy of Technological Sciences and Engineering, *Technology in Australia, 1788–1988.* Melbourne.

Anderson, T. M. and Shea, G. M. (1991). Management and research by the Department of Forestry in Queensland's tropical forests. In Werren and Kershaw.

ANOP Research Services Pty Ltd (1994?). *Community attitudes to environmental issues.* Dept of Environment, Sport and Territories, Canberra.

Australia (1992). *National forest policy statement: a new focus for Australia's forests.*

Australia, Commission of Inquiry into the Lemonthyme and Southern Forests (Helsham inquiry) (c. 1988). *Report of the Commission of Inquiry into the Lemonthyme and Southern Forests.* AGPS, 2 vols.

Australia, Committee of Inquiry into the National Estate (Hope inquiry) (1975). *National Estate.* AGPS.

Australia, Commonwealth Inter-departmental Committee on Regional Forest Agreements (1994). *Discussion paper: Regional forest agreements, a Commonwealth position.*

Australia, Development and Migration Commission (1929). *Report on afforestation and reforestation in Tasmania.* Govt Printer.

Australia, Ecologically Sustainable Working Groups (1991). *Final Report – Forest Use.* AGPS.

Australia, Forestry and Forest Products Industry Council (1987). *Forest industries growth plan.* FAFPIC, Melbourne.

Australia, Department of Primary Industries and Energy, Forestry Branch (1990). *South-east forests of NSW: analysis of wood resource and forest management considerations in the context of the Australian Heritage Commission Act 1975.* Mimeo, 12 pp.

Australia, House of Representatives Select Committee on Wildlife Conservation (1972). *Report.*

Australia, Industry Commission (1993). *Adding further value to Australia's forest products.* AGPS.

Australia, National Occupational Health and Safety Commission (1993). *Estimates of national occupational health and safety statistics, 1991–92.* AGPS.

Australia, National Plantations Advisory Committee (1991). *Integrating forestry and farming: commercial wood production on cleared agricultural land.* Dept of Primary Industries and Energy.

Australia, Resource Assessment Commission (1991). *Forest and Timber Inquiry: Draft report.* AGPS, 2 vols.

Australia, Resource Assessment Commission (1992). *Forest and Timber Inquiry: Final report.* AGPS, 2 vols.

Australia, Resource Assessment Commission (1993). *Ecological impacts of forest use: a survey of completed research.* AGPS (RAC Research Paper no. 9).

Australia, Senate Standing Committee on Science and the Environment (1977). *Woodchips and the environment.* AGPS.

Australia, Working Group on Rainforest Conservation (1985). *Rainforest Conservation in Australia.* Dept of Arts, Heritage and Environment.

Australia, Working Group on the Economic and Environmental Aspects of the Export Woodchip Industry (1975). *Economic and environmental aspects of the export woodchip industry.* AGPS, 2 vols.

Australian Heritage Commission and Department of Conservation and Natural Resources, Victoria (1994). *National Estate values in the Central Highlands of Victoria: draft project report.*

Backhouse, J. (1967) [1843]. *A narrative of a visit to the Australian colonies.* Johnson Reprint Corp., New York.

Barlow, B. A. (1981). The Australian flora: its origin and evolution. In Bureau of Flora and Fauna, *Flora of Australia,* AGPS, Canberra, vol. 1.

Barrère, P., Heisch, R. and Lerat, S. (1962). *La région du Sud-Ouest.* Presses Universitaires de France, Paris.

Batini, F. E. (1987). Planning land use in state forests. In Dargavel and Sheldon.

Bazett, M. D. (1993). *Industrial wood.* Shell International Petroleum Company and World Wide Fund for Nature, London (Shell/WWF Tree Plantation Review, Study no. 9).

Bean, C. (1994). Issues in the 1993 election. *Australian Journal of Political Science.* 29: 136–59.

Beckett, R. 1983. *Axemen, stand by your logs!* Lansdowne Press, Sydney.

Benjamin, L. R. (1959). The challenge of the eucalypts. *Appita.* 13: 90–103.

Benjamin, L. R. (1966). Changing concepts of making newsprint from eucalypts. *Newsprint Log.* June.

Benson, A. J. and Mutton, D. R. (1988). Forestry accounting in Australia. Paper to Austis Conference, Mt Gambier, 19 October 1988.

Black, A. and Breckwoldt, R. (1977). Evolution of systems of national park policy-making in Australia. In D. Mercer (ed.), *Leisure and recreation in Australia,* Sorrett Publishing, Malvern, Vic.

Blackburn, R. (1991). Fin de Siècle: socialism after the crash. In R. Blackburn (ed.), *After the fall: the failure of communism and the future of socialism,* Verso, London.

Boland, D. J., Brooker, M. I. H., Chippendale, G. M., Hall, N., Hyland, B. P. M., Johnston, R. D., Kleinig, D. A., Turner, J. D. (1984). *Forest trees of Australia.* Nelson–CSIRO, Melbourne.

Bolton, G. (1981). *Spoils and spoilers: Australians make their environment, 1788–1980.* Allen and Unwin, Sydney.

Bookchin, M. (1982). *The ecology of freedom: the emergence and dissolution of hierarchy.* Cheshire Books, Palo Alto, Calif.

Boutland, A. (1989). Review paper: forests and Aboriginal society. In Frawley and Semple.

Boutland, A., Byron, N., and Prinsley, R. (1990). *Directory of assistance schemes for trees on farms and rural vegetation.* Greening Australia and Bureau of Rural Resources, Canberra (2nd edn, 1992–93)

Boyden, S. V. (1993). *Culture, nature and the future.* Centre for Human Aspects of Science and Technology, University of Sydney (1993 Templeton Lecture).

Brinkman, R. P. (1990). Advanced forest planning in the Otways. *Australian Forestry.* 53: 290–4.

Bromley, D. W. B. (1991). *Environment and economy: property rights and public policy.* Blackwell, Oxford.

Brown, A. G. (1990). Softwood species selection and breeding of plantations. In Dargavel and Semple.

Brown, L. (1992). Ten years to save the world. *New Scientist.* 1809: 2.

Brundtland Commission, *see* United Nations, Word Commission on Environment and Development.

Budowski, G. (1988). Is sustainable harvest possible in the tropics? *American Forests.* November–December 1988: 34–7.

Byron, R. N. and Douglas, J. J. (1981). *Log prices in Australia: policies, practices and consequences.* BFE Press, Canberra.

Calder, M. (1980). *Big timber country.* Rigby, Adelaide.

Cameron, J. I., and Penna, I. W. (1988). *The wood and the trees: a preliminary economic analysis of a conservation-oriented forest industry strategy.* ACF, Hawthorn, Vic.

Campaign to Save Native Forests (WA) (1975). *The threat to the forests: a submission to the Australian Senate Standing Committee on the Social Environment concerning the impending Manjimup Wood-chip project.* South Perth.

Carron, L. T. (1980). A history of forestry and forest products research in Australia. *Historical Records of Australian Science.* 5: 7–57.

Carron, L. T. (1985). *A history of forestry in Australia.* ANU Press.

Carron, L. T. (1990). A history of plantation policy in Australia. In Dargavel and Semple.

Carron, L. T. (1993). Changing nature of federal–state relations in forestry. In Dargavel and Feary.

Carter, W. G. (1975). Forest management and the export woodchip industry. In Australia, Working Group on the Economic and Environmental Aspects of the Export Woodchip Industry.

Centre for Regional Economic Analysis. (1985). *The export woodchip industry: significance for the Tasmanian economy.* CREA, University of Tasmania.

Chapman, R. J. K. 1992. *Setting agendas and defining problems: the Wesley Vale pulp mill proposal.* Centre of Applied Social Research, Deakin University, Geelong (Deakin Series in Public Policy and Administration no. 3).

Christoff, P., and Blakers, M. (1986). *Jobs in East Gippsland: a transitional economic strategy.* CCV, Melbourne.

Claridge, A. W. (1993). Hypogeal fungi as food resource for mammals in the managed eucalypt forests of South-east Australia. PhD thesis, ANU.

Clay, R. E., Hingston, A. E. K., and Aslin, H. J. (1988). *National parks: perceptions and expectations, a South Australian study.* Centre for Environmental Studies, University of Adelaide (Occasional Paper no. 4).

Cochrane, P. (1980). *Industrialisation and dependence: Australia's road to economic development, 1870–1939.* University of Queensland Press.

Colong Foundation (1983). *How the rainforest was saved: the inside story of the 10 year battle.* Sydney.

Coombs, H. C. (1990). *The return of scarcity: strategies for an economic future.* Cambridge University Press, Melbourne.

Cox, P., and Freeland, J. (1969). *Rude timber buildings in Australia.* Thames and Hudson, London.

Cromer, R. N., Dargavel, J. B., Henderson, V. T., and Nelson, P. F. (1977). More pulpwood from less land. *Appita.* 31(1): 49–54.

CSIRO (1989). *CSIRO paper on the Wesley Vale pulp mill.* 13 March 1989. Mimeo.

Dargavel, J. (1980). The political detection of an Australian forestry perspective. *Australian Forestry.* 43(1): 5–15.

Dargavel, J. (1988a). Early sawmills. In Dargavel (1988c).

Dargavel, J. (1988b). The first timber firm in Van Diemen's Land. In Dargavel (1988c).

Dargavel, J. (ed.) (1988c). *Sawing, Selling and Sons: histories of Australian timber firms.* CRES, ANU.

Dargavel, J. (1990). Social and regional implications of plantation expansion. In Dargavel and Semple.

Dargavel, J. (1992). Incorporating national forests into the new Pacific economic order: processes and consequences. In Dargavel and Tucker.

Dargavel, J., and Feary, S. (eds) (1993). *Australia's ever-changing forests II: proceedings of the Second National Conference on Australian Forest History.* CRES, ANU.

Dargavel, J., and Ferguson, I. S. (1975). Forest use conflicts in Victoria. *Australian Forestry.* 37(3): 215–24.

Dargavel, J., and Semple, N. (eds.) (1990). *Prospects for Australian forest plantations.* CRES, ANU.

Dargavel, J., and Sheldon, G. (eds) (1987). *Prospects for Australian hardwood forests.* CRES, ANU.

Dargavel, J., and Tucker, R. P. (eds) (1992). *Changing Pacific forests: historical perspectives on the forest economy of the Pacific Basin.* Duke University Press.

Dargavel, J., Goddard, J., and Caton, S. (1986–87). *Allocating forest resources in New South Wales (South Australia, Queensland, Tasmania, Victoria and Western Australia): guide to legislation, regulation and administrative practice.* CRES, ANU (CRES Working Papers 1986/32, 1986/38, 1987/1, 1987/2, 1987/3 and 1987/17).

Dargavel, J., Johnston, T., and Boutland, A. (1985). Characteristics of the forestry and logging industry workforces. *Australian Forestry.* 48: 235–41.

Dargavel, J., Parker, B., and Tracey, J. (1992). Australian logging workforce: changes and comparisons in a neglected industry. *Australian Forestry.* 55: 22–8.

Davidson, F. G. and Stewardson, B. R. (1974). *Economics and Australian Industry.* Longman Cheshire, Hawthorn, Vic.

Davis, B. (1986). Tasmania: the political economy of a peripheral state. In B. Head. (ed.), *The politics of development in Australia*, Allen and Unwin, Sydney.

Davis, B. W. (1987). Heritage conservation in Australia: some issues and dilemmas. Seminar paper, Urban Research Unit, Research School of Social Sciences, ANU, 25 May 1987.

Dempsey, K. (1990). *Smalltown: a study of social inequality, cohesion and belonging.* Oxford University Press, Melbourne.

Dempsey, K. (1992). *A man's town: inequity between women and men in rural Australia.* Oxford University Press, Melbourne.

Devall, B. and Sessions, G. (1985). *Deep ecology: living as if nature mattered.* Gibbs M. Smith, Layton, Utah (reprinted in List, 1993).

Dixson, M. (1963). The timber strike of 1929. *Historical Studies Australia and New Zealand.* 10: 479–92.

Dolin, J. E. (1967). The timber industry in Western Australia: a study of organisation and competition. MEc thesis, University of Western Australia.

Donaldson, I., and Donaldson, T. (1985). First sight. In I. Donaldson and T. Donaldson (eds), *Seeing the first Australians.* Allen and Unwin, Sydney.

Dovers, S. R., and Handmer, J. W. (1993). Contradictions in sustainability. *Environmental Conservation.* 20(3): 217–22.

Doyle, T. (1990). Environmental movement power brokers. *Philosophy and Social Action.* 16(8): 37–52.

Dumaresq, D. (1988). Sustainable rural communities: subsistence production within an industrial farming system. In P. Allen and D. Van Dusen (eds), *Global perspectives on agroecology and sustainable agricultural systems,* Agroecology Program, University of California, Santa Cruz.

Dwyer Leslie (1990–93) *Oberon: Rural Community Development Study Reports.* NSW Forestry Commission/State Forests.

East Gippsland Coalition (1985). *East Gippsland proposal for national parks in East Gippsland: a greenprint for the forests.* Melbourne.

Easthope, G., and Holloway, G. (1989). Wilderness as sacred: the Franklin River campaign. In P. Hay, R. Eckersley, and Holloway, G. (eds), *Environmental Politics in Australia and New Zealand.* Centre for Environmental Studies, University of Tasmania (Occasional Paper no. 23).

Eckersley, R. (1992). *Environmentalism and political theory: toward an ecocentric approach.* UCL Press, London.

Ellis, R. C., and Thomas, I. (1989). Pre-settlement and post-settlement vegetational change and probable Aboriginal influences in a highland forest area in Tasmania. In Frawley and Semple.

Elton, H. D. (1976). *Socio-economic study of the timber industry in the Alpine Area: Report to the Forest Industries Resource Management Group.* P.A. Management Consultants Pty Ltd [Melbourne].

ERIN (1993). *Proposed ANZECC list of endangered, vulnerable and extinct Australian species.* Dept of Environment, Canberra (electronic).

ERIN (1994). *Managed area report: the Wet Tropics of Queensland.* Dept of Environment, Canberra (electronic).

Evans, D. J. (1990). Privatisation in New Zealand and its effect on Australia. In Dargavel and Semple.

Evatt Research Centre (1989). *State of siege: renewal or privatisation for Australian State public services?* Pluto Press, Leichhardt, NSW.

FAFPIC (1984). *Stocktake.* Dept of Trade, Business Consultation Unit, Canberra.

FAFPIC (1985). *Strategic plan.* Manufacturing Industries Council.

FAFPIC (1987). *Forest industries growth plan: a submission to the Australian Government.*

FAFPIC (1989). *Potential role of hardwood plantations in supplementing Australia's forest resources: a position statement.* AGPS.

Falconer, R. (1984). Daintree debacle: Australia's great lowland tropical rainforest is in urgent need of help. *National Parks Journal.* 28(1).

Fall, V. G. (n.d.). *The mills of Jarrahdale: a century of achievement.* (n.p.)

Ferguson, I. S. (1985). *Report of the Board of Inquiry into the Timber Industry in Victoria* (Ferguson inquiry). Victorian Govt Printer, 2 vols.

FICA [1988]. *Exposed: the real ACF forests agenda.* South Melbourne.

Fisher, D. E. (1993). The meaning and significance of resource security. In Gardner.

Fitzgerald, F. (1987). Integrated organisations of land management. In Dargavel and Sheldon.

Flinn, D. W., and Turner, J. (1990). Opportunities for increased softwood production through intensive site and nutrient management. In Dargavel and Semple.

Frankel, Sir O. H. (1975). Conservation in perpetuity: ecological and biosphere reserves. In F. Fenner, *A national system of ecological reserves in Australia,* Australian Academy of Science, Canberra (Report no. 19).

Frawley, K. (1991). Past rainforest management in Queensland. In Werren and Kershaw, vol. 3.

Frawley, K. (1994). Evolving visions: environmental management and nature conservation in Australia. In. S. Dovers (ed.), *Australian environmental history: essays and cases,* Oxford University Press, Melbourne.

Frawley, K. J. (1987). *The Maalan Group Settlement, North Queensland, 1954: an historical geography.* Dept of Geography and Oceanography, University College, Australian Defence Force Academy, Campbell, ACT (Monograph Series no. 2).

Frawley, K. J., and Semple, N. (eds) (1989?) *Australia's ever-changing forests* (Proceedings of the First National Conference on Australian Forest History). Dept Geography and Oceanography, Australian Defence Force Academy, Campbell, ACT.

Frost, A. (1994). *Botany Bay mirages: illusions of Australia's convict beginnings.* Melbourne University Press.

Fukuyama, F. (1992). *The end of history and the last man.* Free Press, New York.

Gardner, A. (ed.) (1993). *The challenge of resource security: law and policy.* Federation Press, Annandale, NSW.

Garnaut, R. (1989). *Australia and the Northeast Asian ascendancy.* AGPS.

Gee, H., and J. Fenton (eds) (1978). *The South West book: a Tasmanian wilderness.* ACF, Hawthorn, Vic.

Gibbs, D. M. H. (1992). Economic impacts of the New South Wales rainforest policy decision of 1982 on the timber industry of the Far North Coast. *Australian Forestry.* 55: 80–9.

Goldstein, W. (ed.) (1979). *Australia's 100 years of national parks.* Australian National Parks and Wildlife Service, Sydney.

Green, R. (1981). *Battle for the Franklin.* ACF, Melbourne.

Greenland, H. (1989). The Greens and their political problems. Arena. 89: 23–5.

Griffiths, T. (1992). *Secrets of the forest; discovering the history of Melbourne's Ash Range.* Allen and Unwin, St Leonards, NSW.

Gruen, F., Leslie, A., and Smith, A. P. (1989). *Inquiry into the proposed trial of the Value Adding Utilization System, Central and East Gippsland Forest Management Areas, under the Environmental Effects Act.* Victoria, Dept of Conservation and Environment.

Gunnersen, T. (1977). *The Victorian Alps: a resource base.* FIRM Group, Blackburn, Vic.

Hall, C. M. (1989). The 'worthless lands hypothesis' and Australia's national parks and reserves. In Frawley and Semple.

Hall, M. J. (1990). Barriers and opportunities for growing plantations privately. In Dargavel and Semple.

Hamilton, C. (1993). *The mystic economist.* Willow Park Press, Fyshwick, ACT.

Hamlin, B. F. (1969). Maurice Coleman Davies. *Early Days: Journal and Proceedings of the Royal Western Australian Historical Society.* 6(8): 39–56.

Handa, R. (ed.) (1988) *Forest policy in Japan.* Nippon Ringyo Chosakai, Tokyo.

Hanssen, N. L., and Wigston, D. L. (1989). Approaches to a forest history of the Northern Territory. In Frawley and Semple.

Harris, A. (1953) [1847]. *Settlers and convicts; or recollections of sixteen years in the Australian backwoods; by an emigrant mechanic.* Melbourne University Press.

Harris–Daishowa (Australia) Pty Ltd (1977). *An environmental impact statement for the Eden (NSW) export woodchip operation.*

Harris–Daishowa (Australia) Pty Ltd (1986a). *Draft environmental impact statement for the Eden (NSW) export woodchip operation for the period 1989–2009.*

Harris–Daishowa (Australia) Pty Ltd (1986b). *Environmental impact statement for the Eden (NSW) export woodchip operation for the period 1989–2009.*

Haward, M., and Smith, G. (1990). The 1989 Tasmanian election: the Green Independents consolidate. *Australian Journal of Political Science.* 25: 196–217.

Hawkins, K. M. (1987). Attitudes towards eucalypt clear-cutting among Australian foresters: a social ecological analysis. PhD thesis, Australian National University.

Helman, P. M., Jones, A. D., Pigram, J. J. J., and Smith, J. M. B. (1976). *Wilderness in Australia: Eastern New South Wales and Southern Queensland.* Dept of Geography, University of New England, Armidale, NSW.

Helsham inquiry, *see* Australia, Commission of Inquiry into the Lemonthyme and Southern Forests.

Heyligers, P. C. (1975). Biological and ecological aspects related to the forestry operations of the export woodchip industry. In Australia, Working Group on the Economic and Environmental Aspects of the Export Woodchip Industry.

Heyligers, P. C. (1977). *The natural history of the Tasmanian, Manjimup, and Eden–Bombala woodchip export concession areas.* AGPS (Studies Bureau, Dept of Environment, Housing and Community Development, Report no. 22).

Hinchey, R. (1991). Operation 'Redgum'. Australian Police. October–December: 153–60.

Hindmarsh, W. R. (1983?). The Australian Timber Industry Stabilisation Conference. Mimeo.

Hobsbawm, E. J. (1952). *The age of revolution, 1798–1848.* Mentor, New York.

Holloway, G. (1986). *The Wilderness Society: the transformation of social movement organization.* Dept of Sociology, University of Tasmania (Occasional Paper no. 4).

Hope, G., and Kirkpatrick, J. (1989). The ecological history of Australian forests. In Frawley and Semple.

Hope inquiry, *see* Australia, Commission of Inquiry into the National Estate.

Houghton, N. (1988). Thomas Crowley. In J. Dargavel (1988c).

Howitt, A. W. (1895). *Report on the present and future supply of Victorian Eucalypts, and on the possible establishment of an export trade in timber.* Victorian Govt Printer.

Hudson, I. and Henningham, P. (1986). *Gift of God – friend of man: a story of the timber industry in New South Wales, 1788–1986.* Australian Forest Industries Journal, Sydney.

Hutchins, D. E. (1916). *A discussion of Australian forestry, with special reference to forestry in Western Australia.* WA Govt Printer.

Ishii, Y. (1991). Development of the pulp and paper industry in Japan and the importation of pulpwood. *The current state of Japanese forestry (VII): its problems and future.* Japanese Forest Economic Society, Tokyo.

Izumi, E. (1988). A brief history of Japanese forests and forestry up to the 19th century. In Handa.

Jaakko Pyory (1986). *Competitiveness of Australian forest industries. Report prepared for the Forestry and Forest Industries Council.*

Jackson, R. V. (1977). *Australian economic development in the nineteenth century.* ANU Press.

Jacobs, M. R. (1955). *Growth habits of the eucalypts.* Commonwealth Govt Printer (Forestry and Timber Bureau).

Jarrott, J. K. (1980). The first fifty years: the Romeo Lahey memorial lecture of 1980. *NPA News.* 50(5): 3–19.

Jeffreys, R. B. (1947). A brief historical review of pulp and paper making in Australia. *Appita.* 1: 64–73.

Johns, B. L. (1962). The effect of entry and merger on competition in two related industries. Paper to ANZAAS Conference, Section G, Sydney.

Johnson, D. (1974). *The Alps at the crossroads: the quest for an Alpine national park in Victoria.* VNPA, Melbourne.

Johnson, D. (1982). Conservation issues in Victoria. *Parkwatch.* 131: 27.

Joint Scientific Committee. (1990). *Biological conservation of the south-east forests.* AGPS.

Jolly, N.W. (1920). *Sylviculture.* NSW Govt Printer (Forestry Commission of NSW).

Jones, R. (1969). Fire stick farming. *Australian Natural History.* 16: 224–8.

Jones, R. (1981). The Pedder tragedy: Dick Jones. In Green.

Jones, R. (ed.) (1975). *The vanishing forests? woodchip production and the public interest in Tasmania.* Environmental Law Reform Group, University of Tasmania.

Jones, R., McAllister, I., Denemark, D., and Gow, D. (1993). *Australian election study, 1993: User's guide for the machine-readable data file.* SSDA, ANU (SSDA Study no. 763).

Julius, G. A. (1906). *Western Australian timber tests 1906: the physical characteristics of the hardwood of Western Australia.* WA Govt Printer, 2nd edn.

Kellow, A. (1987). Conflict and intransigence or compromise and incorporation? a dilemma for environmental activists. In *Ecopolitics II: Proceedings of a conference held at the Centre for Environmental Studies,* University of Tasmania, 22–25 May 1987. Centre for Environmental Studies, University of Tasmania.

Kendell, J., and Buivids, E. (1987). *Earth first.* Australian Broadcasting Commission, Sydney.

Kennedy, I. R., and Lodge, J. J. (1976). *Gembrook State Forest: a case study in multiple use management.* Vic. Forests Commission.

Knudtson, P., and Suzuki, D. (1992). *Wisdom of the elders.* Allen and Unwin, Sydney.

Köstler, J. (trans. M. L. Anderson) (1956). *Silviculture.* Oliver and Boyd, Edinburgh.

Kohlsted, S. A. (1983). Australian museums of natural history: public priorities and scientific initiatives in the 19th century. *Historical Records of Australian Science.* 5: 1–19.

Kojima, K., and Ozawa, T. (1984). *Japan's general trading companies: merchants of economic development.* OECD, Paris.

Kumazaki, M. (1988). Japanese economic development and forestry. In Handa.

Langmore, J., and Quiggin. J. (1994). *Work for all: full employment in the nineties.* Melbourne University Press.

Laslett, T. (1875). *Timber and timber trees, native and foreign.* Macmillan, London.

Laughlin, K. (1992). The use of scientists in environmental debate. MLitt thesis, Dept of Geography, ANU.

Lawrence, P. R. (1963). The basis of management planning for integrated utilization in southern Tasmania. *Australian Forestry.* 27(1): 14–26.

LCC, *see* Victoria, Land Conservation Council.

Lennon, J. (1993). Victorian School of Forestry: Demonstration Forest. In Dargavel and Feary.

Lesslie, R., Taylor, D. and Maslen, M. (1993). *National Wilderness Inventory: handbook of principles, procedures and usage.* AHC.

Lewis, H. T. (1982). Fire technology and resource management in Aboriginal North America and Australia. In N. M. Williams and E. S. Hunn (eds), *Resource managers: North American and Australian hunter-gatherers,* American Association for the Advancement of Science, Washington, DC (AAAS Selected Symposium 67).

Lewis, N. B. (1975). *A hundred years of state forestry, South Australia,* 1875–1975. SA Govt Printer (Woods and Forests Dept Bulletin no. 14).

Lewis, N. B. (1988). The Woods and Forests Department of South Australia. In Dargavel (1988c).

Lindenmayer, D. B. (1993). *Wildlife corridors and the mitigation of logging impacts on fauna in wood production forests in South-eastern Australia.* CRES, ANU (Working Paper 1993/4).

Lindenmayer, D. B. (1994). Timber harvesting impacts on wildlife: implications for ecologically sustainable forest use. *Australian Journal of Environmental Management.* 1: 56–68.

Lindenmayer, D. B. (1994). Timber harvesting impacts on wildlife: implications for ecologically sustainable forest use. *Australian Journal of Environmental Management.* 1: 56–68.

Linge, G. J. R. (1979). *Industrial awakening: a geography of Australian manufacturing 1788 to 1890.* ANU Press.

List, P. C. (ed.) (1993). *Radical environmentalism: philosophy and tactics.* Wadsworth, Belmont, Calif.

Loane, B. (1990). Trees on farms and potential of agroforestry. In Dargavel and Semple.

Lothian, J. A. (1994). Attitudes of Australians towards the environment. *Australian Journal of Environmental Management.* 1: 78–99.

Lucas, A. M. (1988). Baron von Mueller: protege turned patron. In R. V. Home, (ed.), *Australian science in the making,* Cambridge University Press, Melbourne.

Ludwig, D., Hilborn, R. and Walters, C. (1993). Uncertainty, resource exploitation, and conservation: lessons from history. *Science.* 260: 17, 36.

Lunney, D. and Moon, C. (1987). The Eden woodchip debate (1969–86). *Search.* 18(1): 15–20.

M'Leod, D. (1868). *Melbourne factories.* Walker, Melbourne.

MacKnight, C. C. (1976). *The voyage to Marege.* Melbourne University Press.

Macpherson, C.B. (1983). The meaning of property. In C. B. Macpherson (ed.), *Property: mainstream and critical positions,* University of Toronto Press.

Maiden, J. H. (1897). Anniversary address. *Journal and Proceedings of the Royal Society of New South Wales.* 31: 67–8.

Maiden, J. H., and Campbell, W. S. (1895–1908). *The flowering plants and ferns of New South Wales with especial reference to their economic value.* NSW Govt Printer (Dept of Mines and Agriculture, Forests Branch).

Major, A. 1991. Is the Eden woodchip dispute a conflict between old and new politics? BA(Hons) thesis, Dept of Political Science, Australian National University.

Mann, M. J. (1958). Integrated utilization in an overmature mixed-species foothills forest. *Australian Forestry.* 22(2): 63–6.

Mann, M. J. (1990). APM Forests plantation projects: the first forty years. In Dargavel and Semple.

Martin, G. (ed.) (1978). *The founding of Australia: the argument about Australia's origins.* Hale and Iremonger, Sydney.

Marx, K. (1954) [1st English edn 1887]. *Capital.* Lawrence and Wishart, London.

Maryborough Chronicle (1988). Wilson Hart & Co. Ltd. In Dargavel (1988c).

Massey, D. B. (1984). *Spatial division of labour: social structures and the geography of production.* Macmillan, London.

Mayer, H. (1964). *The press in Australia.* Lansdowne, Melbourne.

McAllister, I. (1991). *Community attitudes to the environment, forests and forest management in Australia.* Canberra: RAC, Forest and Timber Inquiry (Consultancy Series FTC91/09).

McCormack, R. J., and Sebire, A. (1983). *1980 Logging Industry Census.* CSIRO Division of Forest Research, Canberra (Research Report no. 13, unpubl.).

McCormack, R. J., Roberts, E., and MacArthur, I. (1992). *1992 Logging Industry Census.* CSIRO Division of Forest Research, Canberra (unpubl. internal report).

McCormick, J. (1989). *Reclaiming paradise: the global environmental movement.* Indiana University Press, Bloomington.

McCuaig, A. (1981). *The roots of the woodchip industry in Tasmania: a case study of the origins of Tasmanian Pulp and Forest Holdings Pty Ltd.* Centre for Environmental Studies, University of Tasmania (Project Report 1981/2).

McGrath, K. P. (1957). *The role of utilization in Australian native forest practice.* Commonwealth Govt Printer (Paper for Seventh British Commonwealth Forestry Conference, Australia and New Zealand).

McKinnell, F. H. (1993). Towards 2000 – quo vadis the Institute? *Newsletter of the Institute of Foresters of Australia.* 34(3): 2–8.

McKinty, J. A. (1969). Forestry in East Gippsland. *Proceedings, Royal Society of Victoria.* 82: 129–39.

Mercer, D. (1981a). Trends in recreational participation. In Mercer (1981b).

Mercer, D. (ed.) (1981b). *Outdoor recreation: Australian perspectives.* Sorrett Publishing, Malvern, Vic.

Mercer, D. (1991). *'A question of balance': natural resources conflict issues in Australia.* Federation Press, Annandale.

Merchant, C. (1993). Ecofeminism and feminist theory. In List (1993).

Mills, J. (1986). *The timber people: a history of Bunnings Limited.* Bunnings Ltd, Perth.

Mills, J. (1988). The impact of man on the northern jarrah forests from settlement in 1829 to the Forests Act 1918. In B. Dell, J. J. Havel, and N. Malajczuk (eds), *The jarrah forest: a complex Mediterranean ecosystem,* Kluwer Academic Publishers, Dordrecht.

Mills, J. (1989). The 'teddy bears': a history of the South West Timber Hewers Co-operative Society, Western Australia. In Frawley and Semple.

Milne, C. (1990). Wesley Vale: exposing the inadequacies of ad hoc decision-making. *Canberra Bulletin of Public Administration.* 62: 84–9.

Mitsuda, H. (1984). Impacts of rapid economic growth on the structure of mountain communities: depopulation and disorganization in charcoal-producing mountain villages of Kyoto Prefecture in Japan. In H. K. Steen (ed.), *History of sustained yield forestry: a symposium,* Forest History Society, [Santa Cruz, Calif.].

Mobbs, C. J. (1989). *Nature conservation reserves in Australia* (1988). ANPWS (Occasional Paper no. 19).

Mortlock, W., and Dargavel, J. (1989). *A chronology of forest use and conservation in Victoria with special reference to the Otway region.* CRES, ANU (Otway Project Working Paper 1989/1).

Mosley, G. (1978). Conserving Australia's wilderness; introduction to methods of defining wilderness and the extent of wilderness in Australia. In G. Mosley (ed.), *Australia's wilderness: conservation, progress and plans: proceedings of the first National Wilderness Conference,* ACF, Hawthorn, Vic.

Moulds, F. R. (1991). *The dynamic forest: a history of forestry and forest industries in Victoria.* Lynedoch Publications, Richmond, Vic.

Mueller, F. von (1879). Suggestions on the maintenance, creation and enrichment of forests as applicable to the particular requirements of the Colony of Victoria. In R. W. E. MacIvor, *The chemistry of agriculture.* Stillwell, Melbourne: pp. 272–5.

Mueller, F. von. (1882 et seq.) *Census of Australian plants.* Melbourne.

Mueller, F. von. (1858–82). *Fragmenta phytographiae Australiae.* Melbourne.

Mueller, S. (1984). Know your friends, know your enemies. *Chain Reaction.* 37: 14–15.

Mussen, G. (1919). *The humanizing of industry and commerce, Joseph Fisher Lecture in Commerce.* Hassell, Adelaide.

Naess, A. (1973). The shallow and the deep long-range ecology movement: a summary. *Inquiry.* 16: 95–100. (reprinted in List, 1993).

Native Forests Action Council of Monash University (comp.) (1981). *Pines versus native forests: an annotated bibliography.* Community Research Action Centre, Monash University, Clayton, Vic.

Neave, H. M., Norton, T. W. Claridge, A. W, Cunningham, R. B., and Donnelly, C. F. (1994). *Habitat models for predicting species richness and the distribution of ground-dwelling mammals using data collected from intensive surveys for Potorous longipes in south east New South Wales: first progress report to the NSW National Parks and Wildlife Service.* CRES, ANU.

Nelson, P. J., and Crossland, C. J. (1992). Australian environmental regulations for new kraft pulp mills and the national pulp mill research program. *Appita.* 45(5): 304–6.

Neville, G., and Ellis, R. (1989). IFA Member Opinion Survey – 1988. *Newsletter of the Institute of Foresters of Australia.* 30(1): 15–21.

New South Wales (1870). *The industrial progress of New South Wales, being a report of the Intercolonial Exhibition of 1870 at Sydney.* Govt Printer.

New South Wales, Forestry Commission (1986). *Supplementary document to Harris–Daishowa draft environmental impacts statement for the Eden (NSW) export woodchip operation for the period 1989–2009.*

New South Wales, Pulp and Paper Industry Task Force (1989). *Task force report.* NSW, Dept of State Development.

New South Wales, State Pollution Control Authority (1975). *North Coast Wood Chip Inquiry:* Report of findings.

Northover, C. (1987). Developing national industry policy in a tripartite forum. In Dargavel and Sheldon.

Norton, T. W., and May, S. A. (1993). *Integrated forestry harvesting in Eastern Australia: ecological impacts and priorities for conservation.* CRES, ANU (CRES Working Paper 1993/5).

Norton, T.W., Mackey, B. G., and Lindenmayer, D. B. (1990). Comments on biological and environmental data sets required for the Australian National Forest Inventory. *Australian Forestry.* 53(2): 124–30.

O'Dell, A. C. (1956). *Railways and geography.* Hutchinson, London.

O'Shaughnessy, P. (1986). *Melbourne and Metropolitan Board of Works catchment management policies: a history and analysis of their development.* CRES, ANU (CRES Working Paper 1986/28).

Oates, N. (1982). *Structural change in Mansfield's timber industry: its impact on the community in the eighties.* Shire of Mansfield.

Offe, C. (1985). New social movements: challenging the boundaries of institutional politics. *Social Research.* 52(4): 817–68.

Ord, K., and Dargavel, J. (1989). *Social and economic aspects of forest management in the Otways.* CRES, ANU (Otway Project Working Paper 1989/2).

Osako, M. M. (1983). Forest preservation in Tokugawa Japan. In R. P. Tucker, and J. F. Richards (eds), *Global deforestation and the nineteenth century world economy,* Duke University Press.

Ouchi, Y. (1988). Development of the Forest Plan System (II). In Handa.

Palmer, K. (1991). Aborigines, *values and the environment.* CRES, ANU (Fundamental Questions Paper 7).

Papadakis, E. (1993). *Politics and the environment: the Australian experience.* Allen and Unwin, Sydney.

Papadakis, E. (1994). Development and the environment. *Australian Journal of Political Science.* 29: 68–82.

Parker, B. Dargavel, J., and Ord, K. (1992). Australian forestry workforce: changes, comparisons and productivity. *Australian Forestry.* 55: 65–73.

Parker, P. (1986). *Trade issues and the Tasmanian woodchip debate: export earnings, subsidies and international obligations.* CRES, ANU (CRES Working Paper 1986/22).

Patterson, G. D. (1968). *The tariff in the Australian colonies, 1856–1900.* Cheshire, Melbourne.

Penna, I. (1992). *Japan's Paper Industry: an overview of its structure and market trends.* Chikyu no Tomo/Friends of the Earth Japan, Tokyo.

Pepper, D. (1984). *The roots of modern environmentalism.* Croom Helm, London.

Phillips, B. (1990). *Koalas.* AGPS.

Population Summit of the World's Scientific Academies (1993). Statement. National Academy Press, Washington, DC.

Porritt, J., and Winner, D. (1988). *The coming of the Greens.* Fontana, London.

Potter, L. (1993). The onslaught on the forests in South-East Asia. In H. Brookfield and Y. Byron, (eds), *South-East Asia's Environmental Future: The search for sustainability,* United Nations University Press and Oxford University Press, Tokyo and Kuala Lumpur.

Powell, J. M. (1976). *Environmental management in Australia, 1788–1914: guardians, improvers and profit: an introductory survey.* Oxford University Press, Melbourne.

Powell, J. M. (1988). *An historical geography of modern Australia: the restive fringe.* Cambridge University Press, Melbourne.

Pratten, C. (1983). Pines, policies or promises? *National Parks Journal.* 27(1): 16–18.

Preston, B. J. (1989). *Environmental litigation.* Law Book Co., Sydney.

Purcell, L. (1988). Millars in Western Australia: ninety-nine productive years. In Dargavel (1988c).

Purdie, R., and Cavanagh, M. (1993). Regional assessment of the heritage values of forests. In Dargavel and Feary.

Pusey, M. (1991). *Economic rationalism in Canberra: a nation-building state changes its mind.* Cambridge University Press, Melbourne.

Pybus, C. (1990). Commentary, and Appendix. In C. Pybus (ed.), *The rest of the world is watching,* Pan Macmillan, Sydney.

Pyne, S. J. (1991). *Burning bush: a fires history of Australia.* Henry Holt, New York.

RAC, see Australia, Resource Assessment Commission.

Rainforest Conservation Society of Queensland (1986). *Tropical rainforests of North Queensland: their conservation significance: a report to the Australian Heritage Commission.* AGPS (AHC, Special Australian Heritage Publication Series no. 3).

Rawson, J. (1953). A history of the Australian paper making industry, 1818–1954. M.A. thesis, University of Melbourne.

Reed, J. L. (1954). *Forests of France.* Faber and Faber, London.

Reilley, J. J. (1974). Impact of softwood plantations on rural development. *Australian Forestry.* 37: 142–52.

Richardson, B. (1990). A study of Australian practice pursuant to the World Heritage Convention. *Environment Policy and Law.* 20(4–5): 143–54.

Roberts, S. H. (1968) *History of Australian land settlement, 1788–1920.* Macmillan, South Melbourne.

Robertson, J. R. (1959). The origins of forestry administration in Western Australia: 1829–1929. Mss., Library of Dept of Conservation and Land Management, WA.

Robin, L. (1991). *Building a forest conscience: an historical portrait of the Natural*

Resources Conservation League of Victoria. NRCL, Springvale South, Vic.
Robinson, R.L. (1922). Forestry in the Empire. *Empire Forestry Review.* 1(1): 33.
Roche, M., Dargavel, J. and Mills, J. (1993). Tracking the KTC from kauri to karri to Chatlee. In Dargavel and Feary.
Roche, M. (1992). Privatizing the exotic forest estate: the New Zealand experience. In Dargavel and Tucker.
Rolls, E. (1981). *A million wild acres.* Nelson, Melbourne.
Rosauer, D. (1993). *Conservation of native forests: the changing role of the Australian Heritage Commission.* Paper tabled in the Senate, 13 December 1993.
Routley, R. and Routley, V. (1975) [1973]. *The fight for the forests: the takeover of Australian forests for pines, wood chips and intensive forestry.* Research School of Social Sciences, ANU (3rd edn).
Row, M. (1980). The Huon Timber Company and the Crown: a tale of resource development. *Papers and Proceedings of the Tasmanian Historical Research Association.* 27: 87–102.
Russ, P., and Tanner, L. (1978). *The politics of pollution.* Widescope International, Camberwell, Vic.
Schlich, W. (1982) [1922]. *Manual of forestry: vol. 1: Forest policy in the British Empire.* Bradbury Agnew, London.
Sharp, A. (1968). *The voyages of Abel Janszoon Tasman.* Clarendon, Oxford.
Shepherd, K. R., McKinnell, F. H., Neilsen, W. H., and Squire, R. O. (1990) Silvicultural regimes for pine plantations. In Dargavel and Semple.
Simpfendorfer, K. (1962). Softwood plantations. In Victoria, Forests Commission, *Forests and populations in rural areas.*
Sinclair Knight & Partners. (1988). *Environmental review of an export pulp mill at Wesley Vale, Tasmania.* NBH, Melbourne.
Sinclair, E. K. (1990). *The spreading tree: a history of APM and AMCOR, 1844–1989.* Allen and Unwin, North Sydney.
Sinclair, S. W. (1987). Pacific Basin Economic Handbook. Euromonitor Publications, London.
Sinden, J. (1981). The demand for recreation and tourism in Australia. In Mercer (1981a).
Smith, R. B., and Woodgate, P. W. (1985). Appraisal of fire damage and inventory for timber salvage by remote sensing in mountain ash forests in Victoria. *Australian Forestry.* 48(4): 252–63.
Society of American Foresters (1993). *Task force report on sustaining long-term forest health and productivity.* Bethesda, Maryland.
Specht, R. L., Roe, E. M., and Broughton, V. H. (1974). Conservation of major plant communities in Australia and Papua-New Guinea. *Australian Journal of Botany* Supplement Series 7.
SSDA (1992). *Forests Community Attitudes Survey, Australia, 1990: User's Guide for the machine-readable data file.* SSDA, ANU (SSDA Study no. 632).
Stanley, J. and King, A. (1986). *East Gippsland: an economic evaluation of the Land Conservation Council's proposed recommendations.* National Institute of Economic and Industry Research, Melbourne.
Steane, S. W. (1937). State forestry in Tasmania. *Australian Forestry.* 2(1): 19–24.
Stephens, M., Hansard, A., and Dean, M. (1993). *Competitiveness of selected Australian plantation forests.* ABARE (Research Report 93.10).
Stewart, Hon. D., and McColl, G. (1994). The Resource Assessment Commission: an inside assessment. *Australian Journal of Environmental Management.* 1(1): 12–23.

Streeting, M. and Imber, D. (1991). *The pricing of Australian woodchip exports.* AGPS (RAC Research Paper no. 4).

Stretton, L. E. B. (1939). *Report of the Royal Commission to inquire into the causes and measures taken to prevent the bush fires of January, 1939, and to protect life and property and the measures to be taken to prevent bush fires in Victoria and to protect life and property in the event of future bush fires.* Victorian Govt Printer.

Sugahara, S., and Ouchi, Y. (1988). Development of the Forest Plan System (I). In Handa.

Sydney International Exhibition (1881). *Official record of the Sydney International Exhibition, 1879.* NSW Govt Printer.

Tarlo, K., and Miller, J. (1985). *Forest industry strategy for Tasmania: protecting jobs and forests.* Tasmanian Conservation Trust, Hobart.

Tasmania, Forestry Commission (1978a). *Financial analysis: Forestry Commission operations in concession of TPFH.*

Tasmania, Forestry Commission. (1978b). *The sawmilling industry and the Wesley Vale concession area.*

Tasmanian Trades and Labor Council [1989]. *The ALP: conservation with development, a growing awareness (the unions position).* [FICA, Carlton, Vic.]

Tasmanian Woodchip Export Study Group. (1985a). *Environmental impact statement on Tasmanian woodchip exports beyond 1988.* Tasmanian Govt Printer.

Tasmanian Woodchip Export Study Group (1985b). *Supplement to the draft Environmental impact statement on woodchip exports beyond 1988.* Tasmanian Govt Printer.

Taylor, J. H. (1975). Planning the future use of public land in Victoria. *Australian Forestry.* 37(3): 208–14.

Thompson, P. (1984). *Bob Brown of the Franklin River.* Allen and Unwin, Sydney.

Totman, C. (1989). *The green archipelago: forestry in pre-industrial Japan.* University of California Press.

Totman, C. (1992). Forest products trade in pre-industrial Japan. In Dargavel and Tucker.

Turner, A. (1981). National parks and pressure groups in New South Wales. In Mercer (1981a).

Tyler, M. J. (ed.) (1979). *The status of endangered Australasian wildlife.* Royal Zoological Society of SA, Adelaide.

United Nations, World Commission on Environment and Development (Brundtland Commission) (1987). *Our common future.* Oxford University Press.

Vanclay, J. K. (1992). Lessons from the Queensland rainforests: a century of striving for sustainability. In P. J. Wood, J. K. Vanclay and W. R. W. Mohd (eds) (1993), *The quest for sustainability: 100 years of silviculture and management in the tropics,* Proceedings of the Tropical Silviculture Workshop at the IUFRO Centennial Conference, Berlin, 1–3 September 1992, Forest Research Institute Malaysia, Kuala Lumpur.

Victoria, Department of Conservation, Forests and Lands (1988). *Environmental effects statement: proposed trial of the Value Adding Utilisation System, Central and East Gippsland Forest Management Areas.*

Victoria, Department of Conservation, Forests and Lands (1988?). *Victoria's rainforests: an overview.*

Victoria, Department of Conservation Forests and Lands, East Gippsland Paper Products Development Study (1989). *The process for evaluating the potential for a pulp and paper mill development in East Gippsland.* Mimeo.

Victoria, Department of Conservation and Environment (1990–92). *Statement of resources, uses and values for the Otway Forest Management Area; Proposed forest management plan for the Otway Forest Management Area; Forest management plan for the Otway Forest Management Area.*
Victoria, Government (1986). *Timber industry strategy.* Govt Printer (Government Statement no. 9).
Victoria, Land Conservation Council (1986). *East Gippsland Area Review: proposed recommendations.* LCC, Melbourne.
Victoria, Land Conservation Council (1991). *Wilderness Special Investigation:* proposed recommendations. LCC, Melbourne.
Victoria, Task Force (1983). *Options for future wood production in Victoria.* Minister of Forests.
W. D. Scott & Co. Pty Ltd. (1977). *Environmental Impact Statement for a North Coast Woodchip Export Project as proposed by Sawmillers Exports Pty Ltd.* [n.p.]
Walker, B., and Bridgewater, P. W. (1992). *Scientific aspects of major environmental issues: biodiversity.* AGPS.
Walker, K. J. (1992). Conclusion: the politics of environmental policy. In K. J. Walker (ed.), *Australian environmental policy,* New South Wales University Press, Kensington, NSW.
Warhurst, J. (1990). Australia enters the 1990s. *The Round Table.* 315: 306–14.
Warren, W. H. (1911). *The strength, elasticity and other properties of New South Wales hardwood timbers.* NSW Govt Printer.
Washington, H. (1984). NSW case history – Colo Wilderness, a wilderness won? In J. G. Mosley and J. Messer (eds), *Fighting for wilderness: papers from the ACF's Third National Wilderness Conference 1983,* Fontana–ACF, Sydney.
Watson, A. J., and Cohen, W. E. (1969). Pulping of eucalypts – an historical survey. *Appita.* 22(4): xvii–xxxi.
Watson, I. (1990). *Fighting over the forests.* Allen and Unwin, Sydney.
Wells, A. (1978?). *Australian Timber Industry Stabilisation Conference (AusTIS), Eastern States Timber Industry Stabilisation Conference* (ESTIS). Mimeo, 9 pp.
Werren, G., and Kershaw, P. (eds) (1987, 1991). *The rainforest legacy: Australian National Rainforests Study.* AGPS, 3 vols. (first published by WWF in 1984).
West, J. (1981) [1852]. *The history of Tasmania.* Angus and Robertson, Sydney.
Westoby, J. C. (1987). *The purpose of forests: follies of development.* Blackwell, Oxford.
White, M. E. (1986). *Greening of Gondwana.* Reed, French's Forest, NSW.
Wild, R. (1974). *Bradstow: a study of class and power in a small Australian town.* Angus and Robertson, Sydney.
Wilson, D. (comp.) (1969). *Compendium of Australian forest products statistics 1935–36 to 1966–67.* Forestry and Timber Bureau, Canberra.
Wilson, R. V. (1979). Commercial accounting for a forest service. Paper to 23rd Annual Conference Australian Agricultural Economics Society, Canberra, February 1979.
Windschuttle, K. (1979). *Unemployment.* Penguin, Ringwood, Vic.
Winter, J. W., and Atherton, R. G. (1987). The distribution of rainforest in north-eastern Queensland. In Werren and Kershaw, vol. 1.
Wooten, H. (1987). Conservation objectives. In Dargavel and Sheldon.
Working Group on the Economic and Environmental Aspects of the Export Woodchip Industry (1975). *Economic and Environmental Aspects of the Export Woodchip Industry.* AGPS.

World Commission on Environment and Development, *see* United Nations, World Commission on Environment and Development.

Young, A. K. (1979). *The sogo shosha: Japan's multinational trading companies.* Westview Press, Boulder, Colorado.

Young, O. R. (1981). *Natural resources and the state: the political economy of resource management.* University of California Press.

Young, O. R. (1982). *Resource regimes: natural resources and social institutions,* University of California Press.

Young, P. A. R., and McDonald, W. J. F. (1987). The distribution, composition and status of the rainforests of southern Queensland. In Werren and Kershaw, vol. 1.

INDEX